The Lit Interior

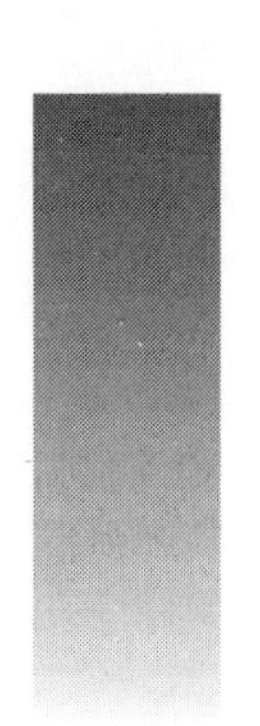

The Lit Interior

William J. Fielder PE, m. IESNA

With contributions by
Frederick H. Jones PhD

OXFORD AUCKLAND BOSTON JOHANNESBURG MELBOURNE NEW DELHI

Architectural Press
An imprint of Butterworth-Heinemann
Linacre House, Jordan Hill, Oxford OX2 8DP
225 Wildwood Avenue, Woburn, MA 01801-2041
A division of Reed Educational and Professional Publishing Ltd

A member of the Reed Elsevier plc group

First published 2001

British Library Cataloguing in Publication Data
Fielder, William J.
The lit interior
1. Interior lighting 2. Lighting, architectural and decorative
I. Title II. Jones, Frederick H. (Frederick Hicks), 1944–
747.9'2

Library of Congress Cataloguing in Publication Data
Fielder, William J.
The lit interior/William J. Fielder; with contributions by Frederick H. Jones.
p. cm.
Includes index.
ISBN 0-7506-4890-2
1. Interior lighting. 2. Electric lighting. 3. Lighting, architectural and decorative. I. Jones, Frederick H. (Frederick Hicks), 1944–
TH7703.F54 2001
621.32'2–dc21 2001053540

ISBN 0 7506 4890 2

For information on all Butterworth-Heinemann publications visit our website at www.bh.com

Composition by Scribe Design, Gillingham, Kent, UK
Printed and bound in Great Britain by Biddles Ltd, *www.biddles.co.uk*

Contents

Preface

This book is intended as a design guide for those individuals in the fields of electrical engineering, architecture, and interior design who will one day design lighting systems for others to build.

The book is organized so that an individual with little or no training in lighting design will become familiar with the basic principles and psychology behind good lighting before design procedures are addressed. Discussions on the process of vision and the properties of light set the stage for exploring the various tools at the designer's disposal for creating and manipulating light to provide a desired effect in an architectural space.

The reader is then led through the conceptual design process, which entails the use of manufacturer's offerings, codes and guidelines for space lighting, as well as calculation methods to predict the performance of a design. The conceptual design is rounded out by exploring methods for powering and controlling a lighting system.

A realistic design problem is begun early in the journey, and is completed, bit-by-bit, as each new concept is explored and applied to the design. Documentation of the design is the final stage of the process, which culminates in a finished set of plans and detailed specifications for the project. A final segment of the book, called 'The Second Time Around', is devoted to retrofitting existing inefficient lighting systems with new, energy-efficient components to improve light quality and reduce the energy consumption of older systems.

Extensive use of the Internet is used throughout the design process. Instructions for downloading and using manufacturer's data, calculation engines, and other tools are included in the text and put to use in the exercises. In the interest of continuity, Internet information for this book is almost exclusively that of Lithonia Lighting Co., a lighting equipment manufacturer. Other manufacturers have similar information available, and the reader is encouraged to search the internet for other favorite sources of information.

William J. Fielder
South Carolina, USA

1 The design medium

The art and science of lighting design is just that, and more: a little artistic flair; some scientific knowledge; and last but not least, a healthy helping of psychology. While every well-done lighting design is attractive, and most provide adequate illumination for the task at hand, the superior design goes the extra mile: it takes into account the effect of the lighted environment on the eye and mind of the human observer. This psychology of the environment is always at play in the relationship of people and architecture, and it can be molded dramatically with effective lighting.

Light can be thought of as a 'building material' much like steel or concrete. Although structural components are needed to enclose a space, the space has no existence for an individual until it is seen and registered in the conscious mind. Light defines space, reveals texture, shows form, indicates scale, separates functions. Good lighting makes a building look and work the way the architect intended at all hours of day and night. It contributes to the character and effective functioning of the space by creating the desired attitude in the mind of the occupant. Change the lighting and the world around us changes.

The actual way the eye–mind combination evaluates light is a complex, dynamic process, which could fill volumes the size of this one. There are, however, some basic principles which bear consideration in the design of lighting systems. In this chapter we will consider both the process of vision, and the effect that light has on our perception of the lighted architecture. You should come away with a better understanding of both the physical and psychological aspects of a lighted environment.

The process of vision

The process of vision can be roughly compared to the operation of a radio or television receiver: there is an antenna, the eye, tuned to

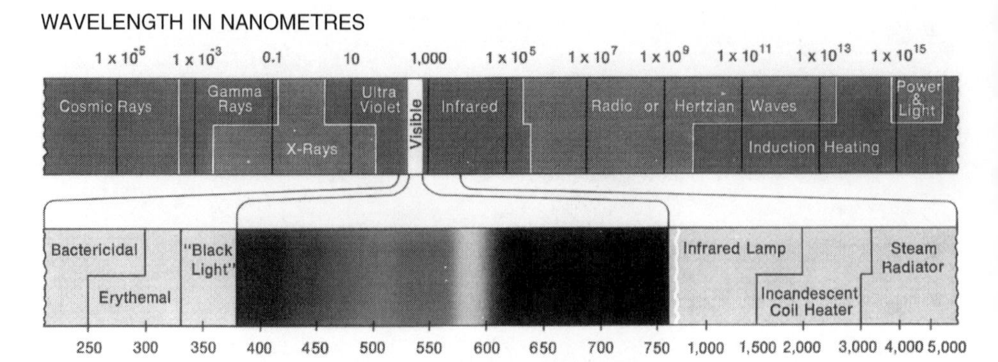

Figure 1.1. The electromagnetic spectrum (source: *Philips Lighting Handbook*).

a specific portion of the electromagnetic spectrum; there's a cable, the optic nerve, connecting the antenna to the decoding device; and then there's the decoding device, the brain, which processes the received information. The eye is tuned to that portion of the electromagnetic spectrum with wavelengths between 380 and 780 nanometers (1 nanometer = 10^{-9} m = 1 thousand millionth of a meter) known as the *visible portion* of the spectrum. Figure 1.1 shows the electromagnetic spectrum, with the visible portion expanded.

As you can see, the visible part of the spectrum covers the wavelengths from ultraviolet, which is commonly associated with skin damage from the sun, to the infrared, which is associated with the heat felt from the sun. This points out the fact that *the shorter the wavelength, the higher the energy* in electromagnetic radiation.

The 'visible' section of the electromagnetic spectrum (see Figure 1.1), when seen simultaneously, appears as white light, such as bright sunlight at noon on a clear day. When white light strikes an object, part of it is reflected, and part is absorbed. For example, a ball which is seen as blue is, in fact, reflecting the blue wavelengths and absorbing all the others.

Our eyes are sensitive to all the wavelengths within the visible spectrum. However, as stated before, they act as 'antennas' to receive reflected light and, like antennas, they are tuned to a specific frequency. In the case of the eye, that frequency lies approximately at the center of the visible spectrum, and has a wavelength of 550 nanometers. This means that the sensitivity of the average eye peaks in the yellow–green portion of the spectrum, and falls off sharply as the limits of the spectrum are approached. Figure 1.2 shows this as a bell curve of eye response relative to light wavelength.

Our eyes not only have to respond to a wide range of wavelengths, but they also must automatically adjust to a constantly varying light intensity. To see how this is accomplished,

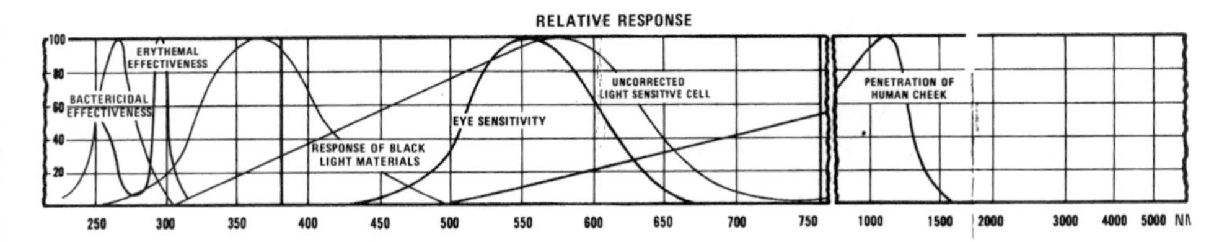

Figure 1.2. Color response of the eye (source: *Philips Lighting Handbook*).

let's take a look at the components of the eye, as shown in Figure 1.3. The 'front end' of the eye acts much like a camera to regulate the incoming light and focus it on the retina. This 'front end' is made up of the *cornea*, the clear outer layer of the eye, and the *pupil*, an opening whose size is constantly being adjusted by the *iris* to compensate for varying light intensity, and the *lens*, which uses the *ciliary muscle* to change its shape to focus the light on a special part of the *retina*, called the *fovea*. The retina contains from 75 to 150 million *rods* and about 7 million *cones*, which make up the actual antennae tuned to the visible spectrum. The rods and cones convert light energy into neural signals that are transmitted to the brain through the optic nerve.

Rods cannot detect lines, points, or colors. They can only detect light and dark tones in an image. Rods are highly sensitive, and they can distinguish outlines of objects in almost complete darkness. Cones are even more sensitive – they detect the lines and points of an image, such as the words you are now reading. Cones also detect

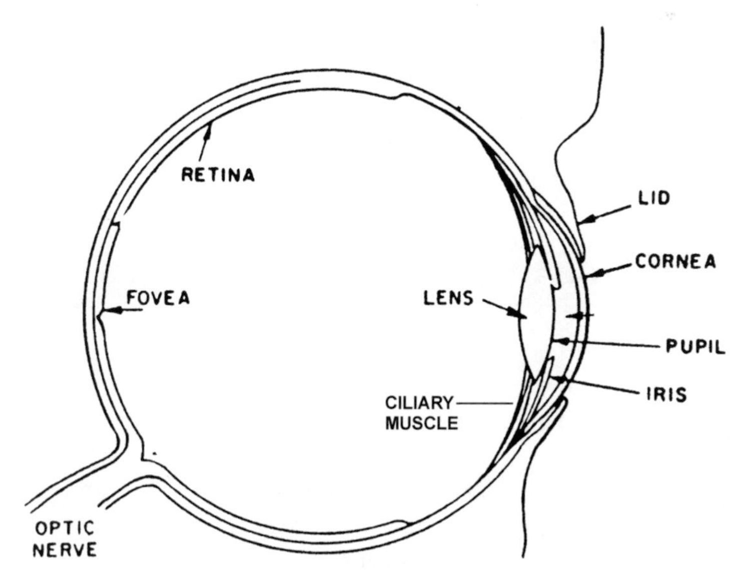

Figure 1.3. Components of the eye (source: F.H. Jones).

color, and there are three types of cones present in the eye: one that is sensitive to the blue–violet end of the spectrum; one sensitive to the yellow–green, or middle of the spectrum; and one sensitive to the red end of the spectrum.

The fovea contains only cones, and provides the optimal reception in brighter light conditions. Muscles controlling the eye work in conjunction with the ciliary muscles controlling the lens to keep the viewed object focused on the fovea. That's why you are moving your eyes while reading this page.

In higher light levels, the cones are the main receptors of light, and the response of the eye to the varying wavelengths of light is as shown in Figure 1.2. In a very low level of light, the cones cease to function, and the sensitivity peak of the eye shifts toward the light with the higher energy wavelengths at the blue end of the spectrum. This is known as the *blue shift*, or *Purkinje effect*, and it is the reason that, under very dim ambient light, the eye will perceive blue light as inordinately bright. This is why police cars in the US have switched from red to blue emergency lights.

As we get older, the components of the eye begin to deteriorate. The ciliary muscles get weaker, the lens loses elasticity, and our ability to focus, particularly on close objects, becomes less. The lens itself yellows with age, which affects color vision, particularly the differentiation between blues and greens. The lens also becomes thicker and less transparent, which results in light scattering and 'night blindness', or extreme sensitivity to glare. The pupil gets smaller, which reduces the overall amount of light which reaches the retina. The result of all this deterioration is that older people need more illumination, larger print, and more contrast in order to see clearly – and to function comfortably.

Now that we know something about the eye, let's take a look at some of the mechanics of those light rays which are constantly bombarding our rods and cones.

Light mechanics

Light travels in a straight line until it strikes a surface. It is then modified by either *transmission, refraction, reflection*, or *absorption.* Figure 1.4 illustrates each of these light modifiers.

Light can also be modified by polarization, diffraction, or interference by other light rays, but these play a very small part in lighting design. For now, let's concentrate on the 'big four', and see how they affect light rays.

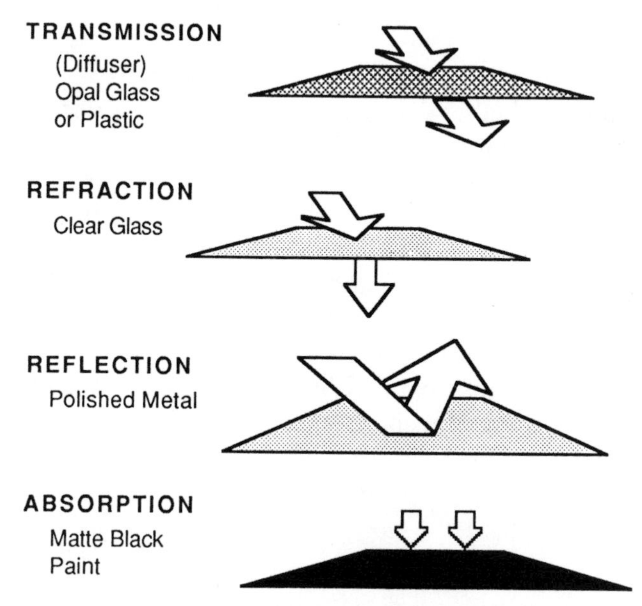

Figure 1.4. Types of light modification (source: F.H. Jones).

1. Transmission

There are three general categories of transmission: *Direct transmission* occurs when light strikes transparent material which can be seen through. These materials absorb almost none of the light in its passage through the material, and do not alter the direction of the light ray. *Spread transmission* occurs with translucent materials in which the light passing through the material emerges in a wider angle than the incident beam, but the general direction of the beam remains the same. *Diffuse transmission* occurs with semi-opaque materials such as opal glass, and the light passing through the material is scattered in all directions. These materials absorb some of the light, and the emerging rays are of less intensity than the transmitted rays. Figure 1.5 illustrates the types of transmission.

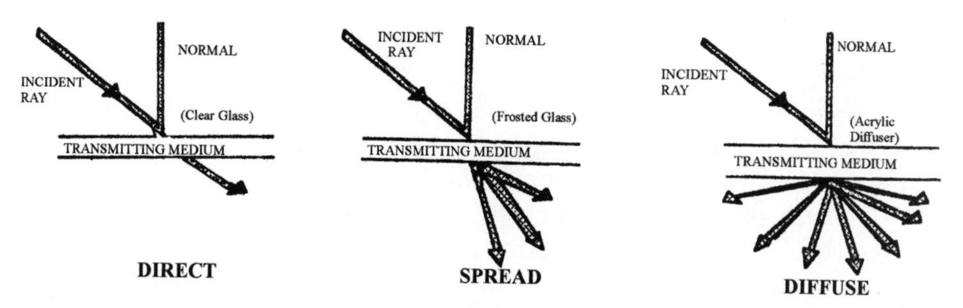

Figure 1.5. Types of light transmission (source: F.H. Jones).

2. Refraction

Refraction occurs when a beam of light is 'bent' as it passes from air to a medium of higher density, or vice versa. This occurs because the speed of the light is slightly lower in the medium of higher density. Two commonly used refractive devices are *prisms* and *lenses*. A prism is made of transparent material which has non-parallel sides. A large prism slows down the various wavelengths of light by different amounts and can be used to divide the light ray into its color components; smaller prisms are used in lighting fixtures to lower brightness or to redirect light into useful zones. Lenses are used to cause parallel light rays to converge or diverge, focusing or spreading the light, as desired. Figure 1.6 illustrates some refractive devices.

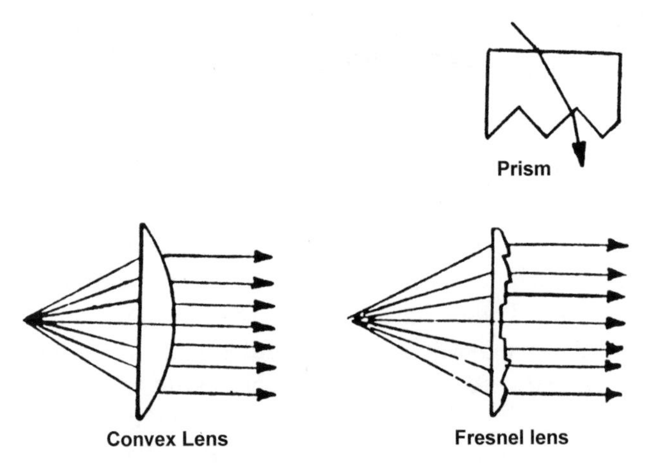

Figure 1.6. Refractive devices (source: F.H. Jones).

3. Reflection

Reflection occurs when light strikes a shiny opaque surface, or any shiny surface at an angle. Reflection can be classified in three general categories: *specular reflection, spread reflection* and *diffuse reflection*. Specular reflection occurs when light strikes a highly polished or mirror surface. The ray of light is reflected, or bounced off the surface at an angle equal to that at which it arrives. Very little of the light is absorbed, and almost all of the incident light leaves the surface at the reflected angle. Spread reflection occurs when a ray of light strikes a polished but granular surface. The reflected rays are spread in diverging angles, due to reflection from the facets of the granular surface. Diffuse reflection occurs when

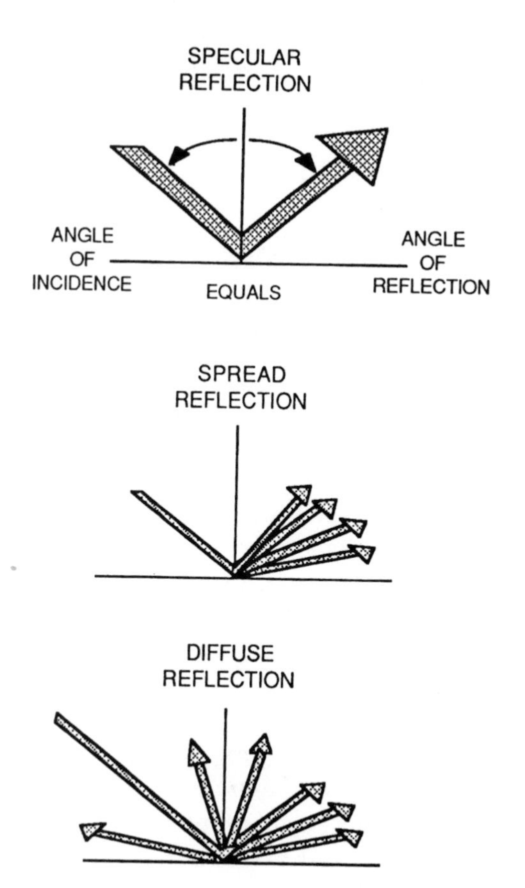

Figure 1.7. Specular, spread, and diffuse reflection (source: F.H. Jones).

the ray of light strikes a reflective opaque but non-polished surface, such as flat white paint. Figure 1.7 shows the types of reflection.

4. Absorption

Absorption occurs when the object struck by the light ray retains the energy of the ray in the form of heat. If you remember the blue ball example, the ball reflects only the blue wavelengths of the incident light, and absorbs all of the others. If the ball were in the sunlight, this energy absorption would heat the ball up. Some surfaces, like flat black paint, absorb nearly all of the incident light rays. These surfaces, such as those of a solar collector panel, tend to get very hot when placed in the sunlight.

With these principles in mind, you can predict how the light itself will behave when used with the various control devices. Now let's look at some of the factors of light which affect the way we see.

Physical factors

In addition to color, the four factors which determine the visibility of an object are: *size, contrast, luminance*, and *time*. Of the four, luminance, or brightness, or the strength of the light falling on the rods and cones, is the underlying dominant factor. Let's look at these factors in more detail.

1. Size

Size is considered because the larger or nearer an object, the easier it is to see. A larger object, of course, reflects more total light, and offers a stronger stimulation of the rods and cones. Also, as we will see in a moment, light adheres to the inverse square law. This means that the strength of the reflected light decreases as the square of the distance between the object and the eye. In other words, the closer the object, the stronger the reflected light.

2. Contrast

Contrast is simply the difference in brightness of an object and its background. Distinct contrast allows the brain to differentiate easily between areas of strong and mild visual stimulation. For example, black words on white paper are read easily, but gray lettering on a slightly lighter gray paper is much harder to interpret.

3. Luminance

Luminance, simply put, is the brightness of an object, or the strength of the light reflected from it. The greater the luminance, the stronger the visual stimulation, and the easier the object is to see.

4. Time

Time refers to how long it takes to see an object clearly. Under the best conditions, it takes slightly less than one-sixteenth of a second for the eye to register an image. In a dim setting, it takes longer. This is especially important where motion is involved, such as in night driving.

Obviously, the luminance of an object, or the quantity of light reflected from it, determines the level of visual stimulation the object provides. Now it is time to look more closely at the mechanics of light quantity, and also to investigate another factor that influences visual acuity, namely, light quality.

Light quantity

In evaluating light quantity, it will be helpful to examine the aforementioned inverse square law, and some of the nomenclature that is used to describe the features of light. Succinctly put, the inverse square law as applied to lighting states that: 'the luminance of an object is directly proportional to the light output of the illuminating source, and inversely proportional to the square of the distance between the source and the object'. At the risk of losing a few of you to the geometry, let's look at Figure 1.8, which graphically illustrates the inverse square law.

Light output from a source is normally expressed in *candlepower*, and light output in a given direction is expressed in *candelas*. The density of light flux radiating from the source is expressed in *lumens*, and the luminance, or light reflected from an object is expressed in *footcandles*. Footcandles has units of lumens per square foot. Figure 1.8 shows a point source of uniform candlepower, having 100 candela in all directions. If we approximate light propagation in a solid angle of 1 steradian and go out a distance of 1 foot from the source, we see that the angle circumscribes an area of 1 square foot. The glossary defines a *lumen* as the flux density generated within 1 steradian by a point source of 1 candela. We have 100 candela in the source of Figure 1.8, so the flux density will be 100 lumens. Using the lumens per square foot definition, we see that the luminance at 1 foot will be 100/1, or 100 footcandles (100 fc). If we go out 2 feet from the source, we see that the area circumscribed by the steradian envelope is now 2 squared, or 4 square feet. Similarly, at 3 feet, the area is 9 square

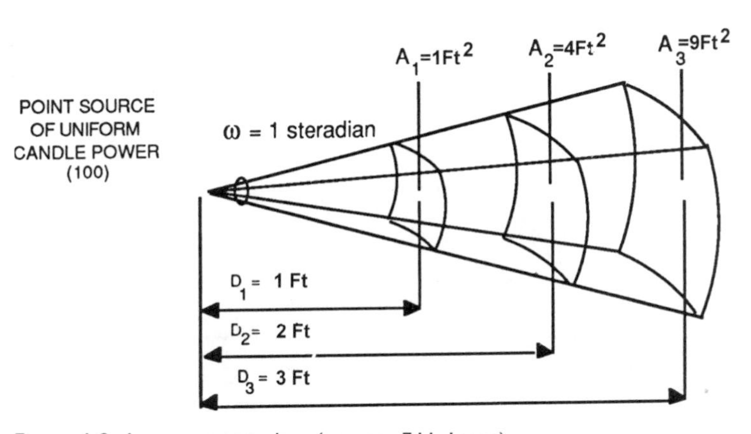

Figure 1.8. Inverse square law (source: F.H. Jones).

feet. Corresponding luminances are 100/4, or 25 fc, and 100/9, or 11 fc, respectively. For the mathematicians among you, this relationship can be expressed mathematically as $I = L/D^2$, where I is illumination in footcandles, L is the luminance of the source in lumens, and D^2 is the square of the distance in feet from the source to the point under examination.

The inverse square law works pretty well in predicting the illumination on a surface from a point source directly above the surface, but what happens when we want to predict the effects of a source that is at an angle to the surface under consideration? We can use an old static mechanics trick and expand the inverse square law to take care of the angle by breaking the angle down into its two components, one parallel to and one normal to the surface, and then discarding the parallel component. Figure 1.9a illustrates this graphically.

Now, if you've ever had a statics course, you'll remember that a force applied to a Point P on a beam at an angle ω from the normal is treated this way, and that the downward component of the force is equal to the total force times the cosine of ω. If you've never had statics, no matter, it still works that way. Taking luminance L as the 'force' of the light, and using the inverse square law, we can say that the illumination I on a point from a source that is at an angle X from being directly above the point, and at a distance D from the point is: $I = L \cos X/D^2$. This is called the *cosine law of incidence*. To get some idea of what this means, look at the light sources above you and all around you. All of these contribute to

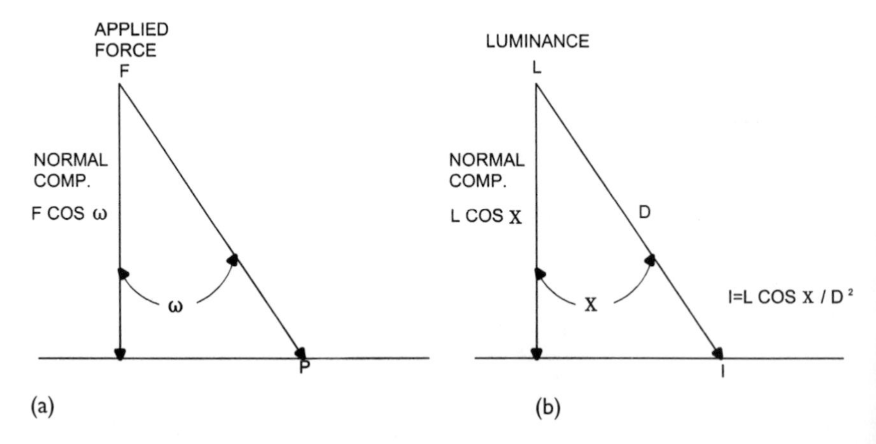

Figure 1.9. Two-component force vector.

the total illumination falling on your desktop. If you have a good calculator, and about a month of free time, you can calculate exactly what that illumination is, using this equation.

Fortunately, we don't have to get bogged down in extensive, tedious calculations of this sort. As we will see in a later chapter, there are plenty of good computer programs out there to perform these calculations for you. It is, however, helpful to know the logic behind the calculations, so that you will be able differentiate between valid output and computer-generated gibberish.

So there you have the factors involved in the quantity of light that illuminates a chosen area. To review in a nutshell, these are the *strength in candlepower* of the illuminating sources; the *distance* those sources are from the area; and the *angle* those sources are from the normal to the surface of the area. Adequate quantity of light, however, doesn't always insure good visibility. The quality of the light is often as important as the quantity.

Light quality

What do we mean by light quality, and what are the factors which contribute to 'good' or 'bad' quality illumination? Simply put, good quality illumination is that which provides a high level of visual comfort, and allows us to view tasks clearly and easily. This affects our psyche in a positive way. On the other hand, poor visual comfort illumination irritates us. The four most important factors affecting visual comfort are *glare, brightness ratio, diffusion*, and *color rendition*. Let's look now at each of these factors in greater detail.

Glare

We've all experienced glare in our everyday lives: bright lighting fixtures located in your field of view, or sunlight coming through a window. This is known as *discomfort glare*, and the degree of discomfort inflicted depends on the number, size, position, and luminance of the glare sources. In interior lighting design, we are primarily concerned with discomfort glare from windows and overhead lighting fixtures. Other forms of glare are *disability glare* and *veiling reflections*. Disability glare obliterates task contrast, and scatters the light within your eye to the point that visibility is reduced to zero. A common example is glare from a glossy magazine page that makes it impossible to read the page. Veiling reflections, such as lighting fixture 'images' on your computer monitor, make it hard to see what is on the screen. The severity of glare in any form is primarily dependent on two factors: the *brightness* and *position* of the source.

Brightness ratio

The brightness ratio is the brightness contrast between the task and the background. This affects the amount of work our eyes have to do in order for us to perform the task. For example, a high brightness task in a low brightness surrounding forces the eye to continually adjust from one light level to the other. Conversely, a low brightness task surrounded by a bright background tends to obscure contrast, and the eye tends to be attracted away from the task. Obviously, a balance between task and background brightness is desirable for effective viewing.

Diffusion

Contrary to the above factors, which affect viewing negatively, diffusion generally improves visual comfort. Diffusion results from light arriving at the task from many different directions. A highly diffuse lighting system will produce no penumbra, or sharply defined shadows. Diffuse lighting is desirable in office areas where computers are in use, in school classrooms, and in library reading areas. Diffuse lighting is accomplished through the use of many low brightness fixtures, or through the use of indirect lighting, where the light is reflected from diffuse surfaces, such as a white ceiling, before reaching the task.

Color rendition

Color of light affects the 'mood', or emotional aspects of a space. It also affects the accuracy with which we perform tasks. We've all, at one time or another, purchased a garment under artificial lighting, only to have it change color when we got it out into the sunlight. That happens because the artificial light source does not contain the full visible spectrum of colors, as does the sunlight. As noted before with the blue ball example, we see only those colors which are reflected from a surface. Obviously, those tasks that involve color discrimination should be lighted by a source that contains as much of the visible spectrum as possible. In other situations, the mood of the environment can be altered by the use of 'warm' or 'cool' colors, high in reds, or blues, respectively.

Unlike light quantity, light quality is subjective in nature, and is not easy to calculate by mathematical formulae. The lighting industry has, however, come up with several methods that the designer can use to evaluate the relative quality of lighting systems. The first of these is *equivalent sphere illumination* (ESI). This is a complicated method of relating illumination of a task on a surface within the design space to that of a task on a surface in the center of a sphere that is equally illuminated throughout. The logic being that the lighted sphere will provide the optimum illumination, and that the space should be designed to match the footcandle requirement of the sphere as closely as possible. For example, if a task requiring 100 fc in the design space was put into the sphere, and the lighting level was adjusted to provide the same task visibility, and that level was 60 fc, then the ESI would be 60 fc. Equivalent sphere illumination takes into account room geometry and reflectance, fixture characteristics, and viewer position. Needless to say, only fixture manufacturers with big computers attempt ESI calculations. Another comparison type system evaluator is the *relative visual performance* (RVP) factor, which is expressed in percentages. The RVP represents the percentage likelihood that a standardized visual task can be performed within the designed lighting system. Age of the viewer, luminance and contrast are all included in RVP calculations. When comparing systems, the one with the higher RVP will provide better light quality. Also expressed in percentages is the *visual comfort probability* (VCP), which is the percent of viewers positioned in a specific location, viewing in a specific direction, who would find the lighting system acceptable in terms of discomfort glare. Visual comfort probability takes into account room geometry and reflectances, fixture number, type and luminance. As with RVP, the higher the VCP, the better the light quality of the installation.

The lighting industry has done a yeoman's job of trying to quantify the factors of light quality so that the above evaluators may be calculated numerically. There are so many non-direct factors involved, however, that these calculations are best left to fixture manufacturers with plenty of time and people, and large computers. Most fixture manufacturers publish some sort of visual comfort data for their fixtures. A good lighting designer is aware of the causes of visual discomfort, and develops an innate 'feel' for which fixtures will perform well where, rather than trying to rely solely on numbers and calculations to provide good light quality.

The psychology of lighting

A seasoned lighting designer can visualize how a given lighting system will look and perform within a space. He also can predict

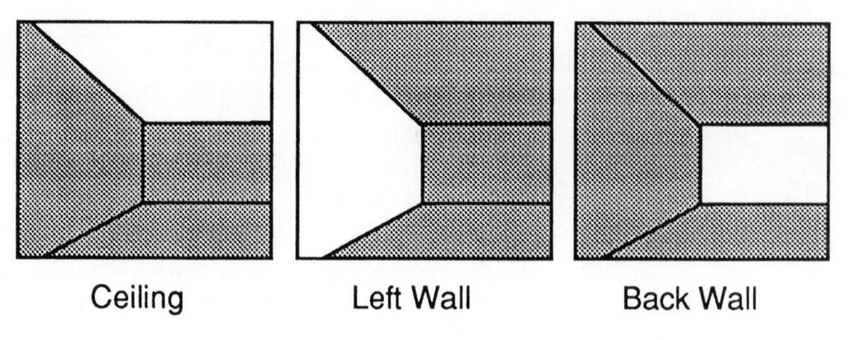

Figure 1.10. Planes of brightness (source: F.H. Jones).

how an observer will react to the system. This insight is gained with experience, of course, but certain basic relationships of light and space and the psyche are always present, and are worth mentioning. The first is the location of the plane of brightness, or the brightest surface in the space. Figure 1.10 illustrates some different planes of brightness.

A ceiling left in shadow creates a secure, intimate, and relaxing 'cave' environment suitable for lounges and casual dining. High brightness on the ceiling creates the bright, efficient, working atmosphere desirable for offices, classrooms, and kitchens. Brightness on the vertical planes draws attention to the walls and expands the space visually, and is appropriate for art galleries, merchandising, and lobbies. Such facilities often also use variations of light intensity on the walls to accentuate a desired feature.

Variations of light intensity form areas of light and shadow, which are desirable if you are trying to create a 'mood' environment, rather than an evenly illuminated workplace. The interplay of light and shadow add variety to a space, and provide visual relief to an otherwise monotonous environment. Scallops on a wall from downlights, shadows on the ceiling from uplights, or highlights from accent lighting create areas of visual interest, and can draw attention to a desired area or object. The designer must be careful not to overdo it, though, because too many lighting effects in one space have roughly the same visual effect that too many sidebars, colors, and font styles do to a magazine page: the original design intent is obscured or obliterated.

It is always best to work with the architect from the outset of a project to get in tune with the flavor or mood that he or she is trying to create in a space. Architectural features can be modeled through the use of shadows, as can objects within the space. A three-dimensional object lighted directly from in front will appear

Figure 1.11. Effects of shadows (source: F.H. Jones).

flat, but when lighted from an angle, will assume depth and roundness. We all remember the 'Frankenstein flash' from photography class, where the hand-held flash is placed beneath the chin of the subject. The resulting photo looks like a Boris Karloff publicity shot. This happens because of the striking contrast between the brightly lit and deeply shadowed facial features. The same effect can be achieved in architectural spaces through the use of uplights, downlights, and lighting from the side. Figure 1.11 illustrates modeling through the use of shadows.

Areas of brightness can also be used to create mood, or accent architectural features. Small pinpoints of brightness from tiny lamps or reflections add sparkle and glitter to a space, which enhances the gaiety, elegance, and festivity of the space. You can try this yourself at home with a string of clear Christmas mini-lights. Wrap that ficus tree in the corner with the lights, and plug them in. Note the overall effect. Or string the lights around an architectural feature, such as an arched doorway. You'll get the idea. The watchword here is the same as with light and shadow, don't overdo it – and keep some background light. Without sufficient background

Figure 1.12. Sparkle and glitter (source: F.H. Jones).

lighting to soften the contrast, glitter can become glare. The lighting is used to accent the architecture – not overpower it. Figure 1.12 shows an effective use of sparkle and glitter.

Summary

We now know the nature of light, how the eye responds to light, and how light behaves when subjected to various control mediums. We've also explored methods for quantifying the intensity of light, and some methods for assessing the visual comfort of a lighting system. All of this knowledge will come into play in every lighting system that you design.

We have also looked at some of the artistic aspects of lighting design. Although these will probably not be a majority of your total design effort (unless you specialize in artistic lighting design), this is where you can use your creativity to create your own unique lighted environment.

In the next chapter we will explore some of the tools that you have at your disposal for creating a lighting system. But first, let's see how much you've retained from this chapter. . .

Exercises

1. What is the wavelength range of the visible portion of the electromagnetic spectrum?
2. Which light rays contain the most energy?
   a. Ultraviolet
   b. Infrared
   c. Blue–Green
3. When we see an object, we are actually seeing
   a. Light striking the object
   b. Light reflected from the object
   c. Light being diffracted by the object
4. We see colors by using the eye's
   a. Rods
   b. Cones
   c. Iris
5. What are four things that happen to our eyes as we age?
6. Name four ways that a light ray can be modified.
7. What is the purpose of a lens?
8. What are the three types of reflection?
9. What are the most important factors determining the visibility of an object?
10. In 31 words or less, parrot back the definition of the inverse square law as it applies to lighting.
11. Problem: A point source of 200 lumens, located directly above a surface, and 2 feet from it, will produce how many footcandles on the surface?
12. What law would we invoke to calculate illumination by a source at an angle from the normal to the lighted surface?
13. What is considered 'good quality' illumination?
14. What is disability glare? Why do we call it that?

15. What determines brightness ratio?
16. Where would diffuse lighting be used?
17. What determines the perceived color of an object?
18. What does a visual comfort probability (VCP) of 70 tell us about a lighting system?
19. Which ceiling brightness would we try to obtain in an upscale dining space?
    a. High
    b. Medium
    c. Low
20. True or false: To add spice to a space, we always try to use as many lighting effects as the budget will allow.
21. What methods would you use to achieve effective shadows?

2 The design tools

Just as an artist has paints and canvas, and a sculptor has chisels and marble, a lighting designer has certain tools with which to create a lighted environment. There is a wide variety of lighting sources, or lamps, available to provide the desired light intensity and color, and there is an almost unlimited choice of lighting fixtures (called *luminaires* from here on out) to shape and place the light in just the right pattern, and there is a multitude of control systems to make those luminaires behave as the designer intends. Like the artist, whose canvas has finite limits, the lighting designer has energy, task, and code requirements that establish the boundaries of the design. To aid in meeting these requirements, the designer has several calculation techniques, both manual and computerized, available to help predict the adequacy and long-term performance of the design.

As with any endeavor, the lighting designer will develop skill and technique with experience. Until that experience is gained, there are a number of 'tried and true' universal lighting techniques that may be used to solve a variety of design problems. Some of these techniques are included in this chapter.

Understanding this powerful arsenal of tools will enable the designer to specify a superior lighting system for any space with confidence, be it a warehouse or a specialty retail shop.

Let's now take a closer look at these tools.

Lamps – the light source

Lamp theory

Artificial light, meaning something other than sunlight, may be produced either chemically, mechanically, or electrically. Chemical light is used for very special applications, such as the emergency wands that you see in sporting goods stores; mechanical (flame –

gas lamps) is primarily used to create atmosphere. By far, the greatest producer of usable artificial light is electricity. The electrically driven light source is what we will examine here.

The two most important methods by which we can produce light from electricity are *incandescence* and *photoluminescence*. These are big words used to describe the processes by which the lamps that we use every day produce light. Let's see how each of these processes works.

Incandescence

Incandescence is the visible radiation produced by a body at a high temperature. In incandescent lamps, the temperature is created by passing an electric current through a wire filament which has a finite resistance to current flow. The result is called Joule heating, and is proportional to the resistance of the filament, and to the square of the current flowing. As the filament gets hot, it glows, first red, and then white. The white light produced by a glowing filament contains all of the colors of the visible spectrum to some degree, and therefore produces a *continuous spectrum* of light. In incandescence this spectrum is higher in intensity toward the longer wavelengths, or the 'red' end of the visible spectrum, but the actual intensities of the various wavelengths depend upon the temperature of the filament. Figure 2.1 shows the spectral energy distribution of a tungsten filament at 3000 K.

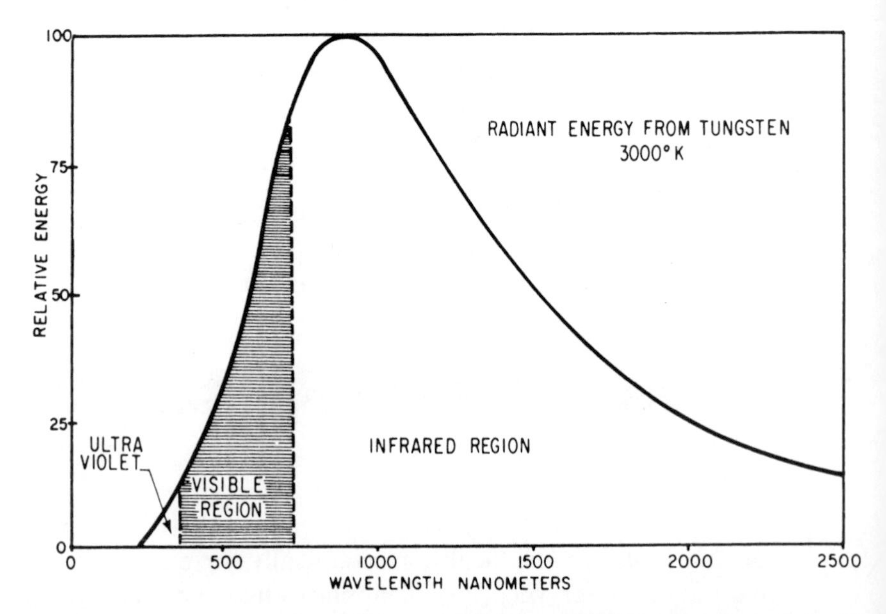

Figure 2.1. Spectral distribution of a tungsten filament at 3000 K (source: Osram/Sylvania).

Photoluminescence

Photoluminescence is what happens when *neutral gas atoms collide with electrons in an electric arc discharge*. Some of the energy of the collision is released as visible radiation, and some as ultraviolet (UV) radiation with wavelengths below the visible spectrum. The energy is released at specific wavelengths which are dependent on the chemical makeup of the gas. The resulting spectral distribution is a discontinuous series of 'spikes', rather than the smooth distribution curve of the incandescent source. This is illustrated in Figure 2.2, which is the spectral energy distribution for a mercury vapor lamp.

Figures 2.1 and 2.2 are what is called *spectral energy distribution curves*. They show what wavelengths, and hence what colors of light are produced by the source under examination. These curves are a good indicator of the color-rendering abilities of the source.

Remember the blue ball example from Chapter 1? OK, let's assume that the color of the ball is about 440 nanometers in wavelength.

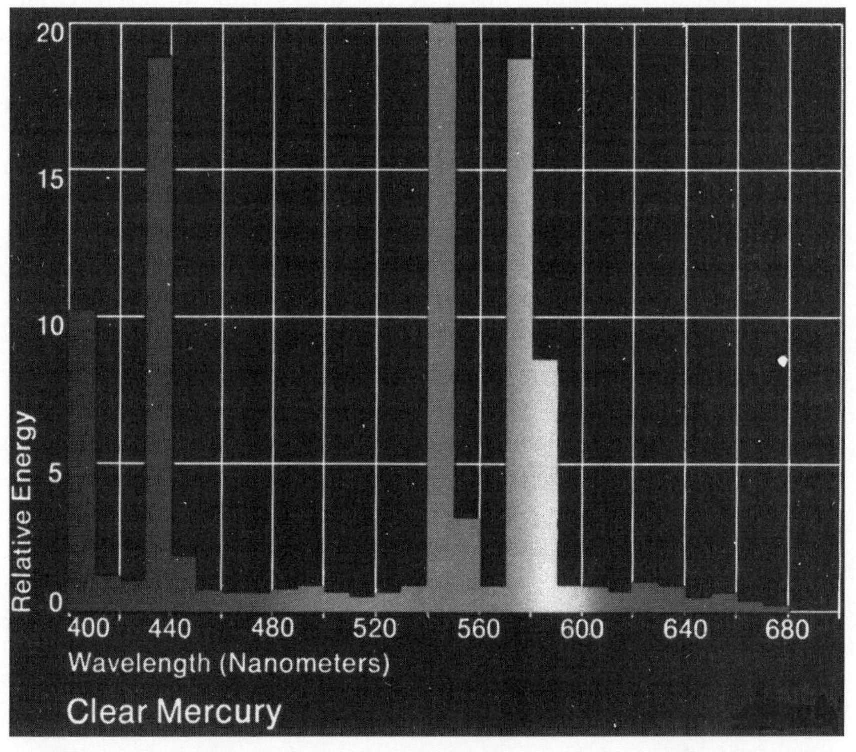

Figure 2.2. Spectral distribution of a clear mercury vapor lamp (source: Westinghouse).

What would the ball look like if it were illuminated by the mercury vapor source in Figure 2.2? Since the ball's color is contained in the spectrum of the source, the ball will reflect the 440 nanometer wavelength light, and appear . . . bright blue. What would a red ball of 700 nanometers wavelength look like? Right – brownish-black! Since very little 650 nanometers wavelength light is in the mercury vapor spectrum, the ball will absorb almost everything. Now you know why meat markets don't use mercury vapor lamps.

Color temperature

It is common to speak of 'cool' colors such as blue and green, and 'warm' colors, such as red and orange. These colors, of course, are the result of the spectral distribution of the light source illuminating them. Light sources, then, can be called cool or warm, or in-between, dependent upon their spectral distribution. Lighting designers find it useful to be able to assign temperature numbers to define the degree of coolness or warmth of a source.

In correlating color to temperature, it is helpful to think of the old time blacksmith heating a piece of iron. As the iron started to heat up, it would glow a deep red. As it got hotter, it would become bright red, and finally, white hot. The easiest way to describe the color of the glowing metal was to give its temperature, because any two glowing pieces of the same metal having the same temperature would always have the same color.

The tungsten of an incandescent lamp filament behaves in the same way, and the color of its emitted light is always directly related to its temperature in Kelvin degrees. The whiter the light, the higher the temperature. The color temperature of a household incandescent lamp, for example, is a little less than 3000 K. Strictly speaking, color temperature applies only to incandescent sources and to natural sources such as the sun and the sky.

When it comes to photoluminescent sources such as fluorescent and gaseous discharge lamps, the correct term to use is *correlated color temperature*. When a fluorescent lamp is said to have a correlated color temperature of 3000 K, that means that its color looks more like that of a 3000 K incandescent source than that of an incandescent source of any other temperature. This does not mean that it will illuminate colored objects the same way a 3000 K incandescent source will, however. Figure 2.3 shows the spectral energy distribution curves for several popular fluorescent lamps.

As shown in Figure 2.3, fluorescent lamps do not have the smooth continuous output curve of an incandescent lamp, so its color temperature is only an apparent one. Figure 2.3 illustrates

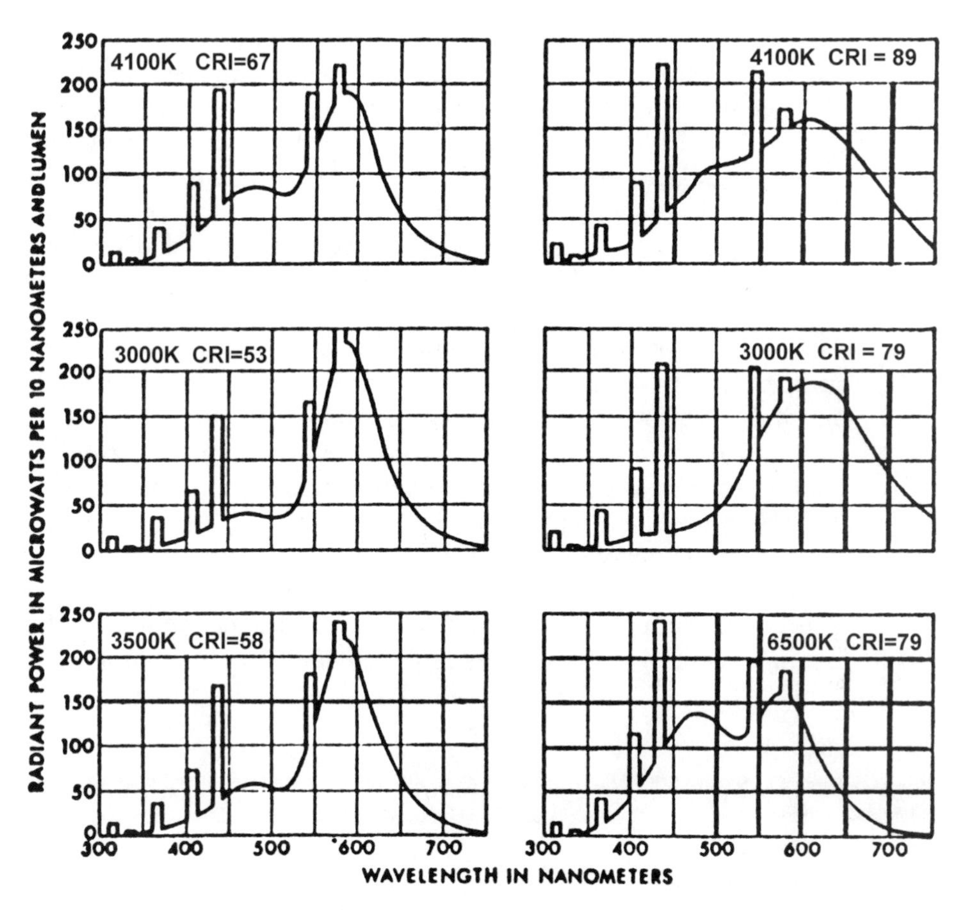

Figure 2.3. Spectral distribution of fluorescent lamps (source: F.H. Jones).

that two 4100 K fluorescent lamps can have very different spectral distributions. In fact, with the right mix of phosphors, it would be possible to make a fluorescent lamp that had most of its output in only two narrow peaks of light, one blue and one yellow, and balance the colors so that the lamp had a correlated color temperature of around 3000 K. The color performance of the lamp would be very poor, however. A red object would appear brown under this lamp, while it would appear normal under a 3000 K fluorescent source with an spectral distribution containing reds.

Color rendering index

Lamp manufacturers, as well as lighting designers, find it useful to be able to compare how well the different sources render colors, so in 1965 an international panel of experts, the Commission Internationale de la Eclairage (CIE) devised an extremely complicated method of testing to assign a *color rendering index* (CRI) number to each lamp on the market. Fortunately, doing the mathematics falls to the manufacturer, and all the designer has to remember is that the higher the CRI, on a scale of 0–100, the better that lamp will render colors. Figure 2.3 lists the CRI for each lamp shown. It is seen that lamps of the same color temperature can have very different CRI ratings, depending on the colors contained in its spectral distribution.

In general, the higher the CRI, the better a fluorescent or discharge source compares with a natural source at the *same* correlated color temperature. It has been said that correlated color temperature is what the lamp is trying to be, and its color rendering index shows how well it is succeeding.

Lamp manufacturers have devoted large amounts of time and money to developing exotic phosphor coatings for the inside of photoluminescent lamps in order to improve CRI. As a result, some fluorescent lamps with a CRI of over 90 are readily available.

Let's take a closer look now at the three major types of light sources, or lamps, which you, as a lighting designer, will use. These are *incandescent, fluorescent,* and *high intensity discharge (HID)* lamps.

Lamps – major types

Incandescent lamps

Standard incandescent lamps

The standard incandescent lamp is the oldest electric lighting technology still available today. Although relatively short lived (750–3000 hours), and fairly inefficient (6–24 lumens per watt), incandescent lamps are widely used in residential and other applications. This is primarily because they are cheap, easy to obtain, work in inexpensive fixtures, have good color rendition, and perform well in low ambient temperatures. Standard incandescent lamps are still constructed today much the same way they were when they were originally introduced: two lead-in wires are attached to a metal screw base, and are insulated from each other by a glass stem. A tungsten filament is attached between the two lead-in wires, and a glass envelope, called the bulb, surrounds the filament structure and is attached to the base. The bulb is evacuated, then filled with an inert gas to prevent the filament from burning up. The envelope may be frosted to diffuse the light emitted from the filament. Figure 2.4 illustrates the construction of the incandescent lamp.

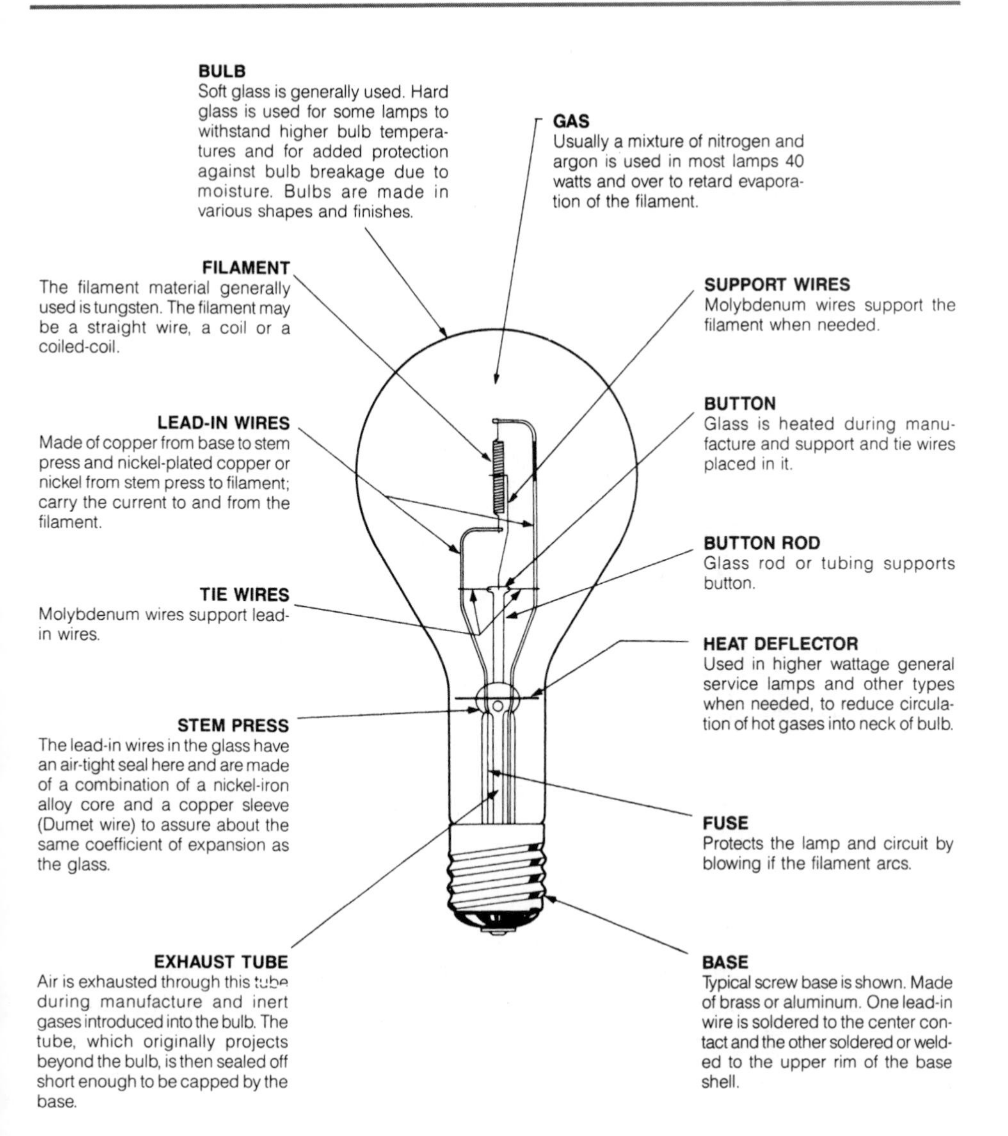

Figure 2.4. Incandescent lamp construction (source: Osram/Sylvania).

In operation, electric current is passed through the filament, which heats up, and produces visible light. The amount of current flowing in the filament determines the brightness and the color of the lamp, from dull red to bright white. An incandescent lamp is easily dimmed by simply reducing the current through the lamp. This is easy to accomplish using a rheostat, or variable resistor, in the lamp circuit.

As you can see from Figure 2.1, the light produced by a tungsten filament is strongest in the long wavelengths, or red region, of the visible spectrum. Incandescent sources can therefore be considered

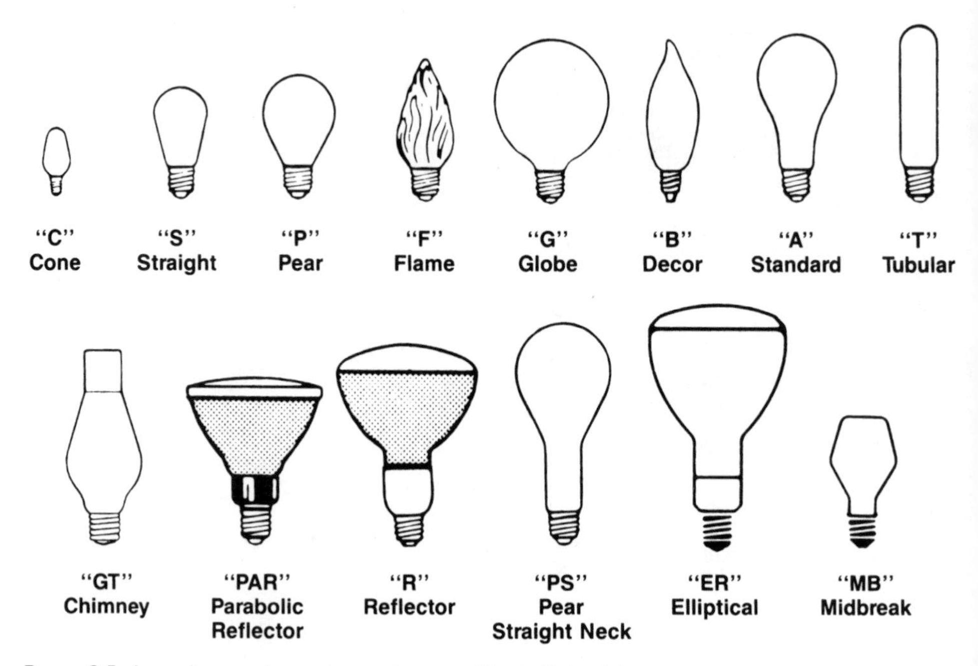

Figure 2.5. Incandescent lamp shapes (source: Osram/Sylvania).

'warm' sources, because reflection from an object will be strongest in the red region.

Standard incandescent lamps are manufactured in a large variety of shapes and sizes. Figure 2.5 shows some of the more common ones.

Each shape of incandescent lamp is assigned a letter, or letters, and each size is assigned a number. The letters loosely describe the lamp, and the number tells the diameter of the lamp in eighths of an inch. For example, the A-19 lamp in Figure 2.5 is of Arbitrary (or Apple) shape, and is 19/8, or 2 3/8 in. (6 cm) in diameter. Some combinations describe the function of the lamp as well as the size, like the PAR-38, which has a *p*arabolic *a*luminized *r*eflector coating inside the glass envelope, and can be bought in either flood or spot light configuration. It is 38/8, or 4 3/4 in. (12 cm) in diameter.

In the years since the invention of the incandescent lamp, lamp manufacturers have done extensive R & D to improve the life and the lumen output of the lamps. This work has been directed toward both the envelope and the gas which fills the envelope.

Tungsten halogen lamps

The tungsten halogen lamp is one result of this effort. The halogen fill gas prevents the lamp envelope from darkening as the lamp ages by carrying vaporized tungsten back to the filament, so more light is produced, and the life of the lamp is extended. The envelope itself is small, and made of quartz, instead of glass like regular incandescent lamps. This combination allows the tungsten halogen lamp to operate at a much higher temperature than a standard incandescent lamp, resulting in a higher efficacy, or lumen per watt, output. Since the lamp is hotter than a standard incandescent, the light output is whiter, and the color rendition is more evenly balanced than the red-rich output of the standard lamp. Tungsten halogen lamps operate with a color temperature of 3000–3300 K and have a CRI of 100. Two types of tungsten halogen lamps are available: the line voltage; and the low voltage lamp, and each type is available in a variety of shapes and sizes.

Infrared reflecting lamps

One version of the tungsten halogen lamp, the infrared (IR) reflecting lamp, uses a dichroic coating that reflects infrared back into the lamp, and passes visible light as output. This serves two functions: it reduces the lamp's energy wastage to heat; and it helps to heat the filament to a higher temperature. A variation of this coating is used in the low voltage MR-16 lamp to produce a 'cool' brilliant white light beam. This lamp gets extremely hot, and cannot be used in an indoor luminaire without protective shielding.

Incandescent lamp benefits

Some of the more prominent benefits of incandescent lamps are:

1. inexpensive;
2. high CRI (good color rendition);
3. operates in inexpensive fixtures;
4. easily dimmed;
5. instant on–off;
6. insensitive to ambient temperature;
7. available with a variety of built-in reflectors;
8. available in many wattage and colors.

Incandescent lamp drawbacks

Some of the drawbacks of incandescent lamps are:

1. short life (750–3000 hours);
2. poor over-voltage tolerance;
3. low lumen output per input watt (efficacy);
4. high heat generation (90% of input wattage goes to heat).

Incandescent lamp uses

Incandescent lamps are generally best suited for use in:

1. dimmable systems;
2. accent and specialty retail lighting;
3. outdoor convenience and decorative lighting systems;
4. mood lighting.

Fluorescent lamps

Fluorescent lamps are the most commonly used commercial light source in the US, and perhaps the world. In the US, about 75% of the commercial space is illuminated by a fluorescent source. This is because fluorescent lamps are relatively cheap, long lived, have a high light output to watts input ratio (efficacy), and are designed for use in fixtures which fit into the architectural schemes of commercial structures.

Let's look at how a fluorescent lamp is built, and see how it operates.

Fluorescent lamps are built using a tubular glass envelope coated on the inside with a mix of phosphors. Inert gas and a small amount of mercury is introduced into the tube to provide the atoms for photoluminescence. The tube is slightly pressurized, and the ends of the tube are capped with electrodes, which contain a cathode to generate an arc.

In operation, when the arc is struck between the electrodes at each end of the tube, the mercury vaporizes, and the electric arc colliding with the atoms of mercury vapor produces UV light along the length of the tube. This high-energy light strikes the phosphor coating, and imparts energy to it, which causes the phosphor to fluoresce, or produce light. The chemical makeup of the phosphor compound determines the color of the light produced.

If this process were allowed to go unchecked, the current flow in the arc would continue to increase until the lamp overheated and destroyed itself. To regulate the current flow, a ballast transformer must be used in conjunction with the lamp. The ballast transformer also serves to start the arc in instant start lamps, and to provide coil voltage in rapid start lamps. Since it must generate higher voltages, the ballast transformer used with instant start lamps is larger than that used with rapid start lamps. A larger ballast is also used for the high output (HO) and very high output (VHO) lamps, which are constructed to allow a larger current flow in the arc.

Fluorescent lamps are available in correlated color temperatures of 2700 K, which approximates incandescent lighting, all the way up to 6500 K, which approximates daylight. The lower temperature lamps, up to 3400 K, are said to be 'warm' lamps; the mid-range lamps, from 3500 to 4000 K, are considered 'natural', neither warm nor cool; and lamps having temperatures of 4100 K and above, are called 'cool' in color. The lamps are available in color rendering indices (CRI) from 57 to above 90. In general, the higher the CRI, the more expensive the lamp.

The two most popular types of fluorescent lamps, the rapid start and the instant start, utilize different methods of arc starting.

Rapid start fluorescent lamps

The rapid start lamp uses a cathode consisting of a coiled tungsten wire coated with an emission material. When voltage is applied across the coil, it heats up, and the coating emits electrons. These electrons produce an arc in the inert gas. This type of lamp necessarily has two pins in each electrode to power the coil.

Instant start fluorescent lamps

The instant start lamp uses a ballast transformer to boost the voltage up to a sufficiently high level to strike the arc directly. This only requires a single pin in each electrode, although many lamp manufacturers use the two-pin configuration, and connect the pins together. Figure 2.6 shows the components of the rapid start fluorescent lamp.

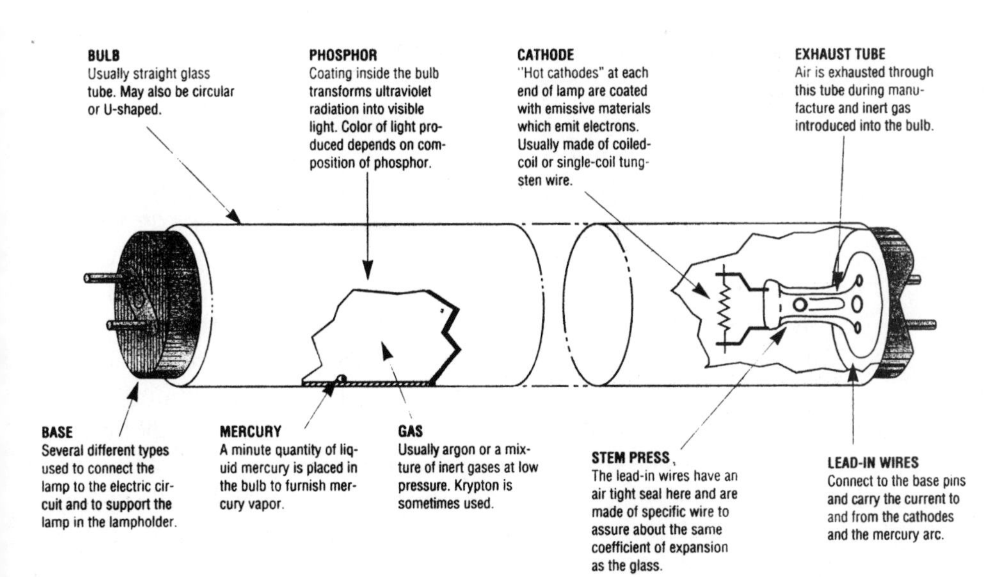

Figure 2.6. Fluorescent lamp construction (source: Osram/Sylvania).

High output and very high output fluorescent lamps

High output (HO) and very high output (VHO) fluorescent lamps are constructed in the same manner as standard fluorescent lamps, except that heavier components are used to allow higher than normal current flow within the arc. As a result of this higher current flow, the arc–electron collisions are more violent, and the lamp produces more lumens than a standard lamp. The cost of this extra output is the increased energy required to operate the lamp. This cost is often justified if higher mounting heights are required by the application.

Compact fluorescent lamps

Compact fluorescent lamps have the same components, and operate the same way that the large tubular lamps do. The glass envelopes of the compact lamps are usually bent into a 'U' shape to offer more surface area in a small space. There are compact fluorescent lamps available that will serve as low-energy substitutes for incandescent lamps in all applications up to about 150 W. The general rule of thumb is that 1 input watt of compact fluorescent energy produces as much light output as 4 input watts of incandescent energy.

Other fluorescent lamps

There are also fluorescent lamps available which have *no* electrodes, but instead control the arc through inductive coupling. The Osram–Sylvania 'Icetron'™ is one such lamp. Since there are no electrodes to deteriorate, these lamps have an extremely long life – about 100 000 hours on average. These lamps are relatively expensive, and are used primarily where lamp replacement is difficult, for example, in a theater with 40 foot (12 m) high ceilings.

Fluorescent lamps are described in the manufacturer's literature in much the same way that incandescent lamps are: the first letter in their designator is either 'F', for fluorescent, or 'CF' for compact fluorescent. The second letter in the designator is used to describe manufacturer's information about the lamp. For example, 'FB' would describe a U-shaped, or 'bent' tube. One lamp manufacturer, Osram–Sylvania calls their T8 energy saving lamps 'Octron', so the designator would be 'FO' for a Sylvania T8 lamp. Absence of a letter in this position indicates a standard lamp. The first number in the designator is usually the wattage of the lamp, with a few exceptions: the number '48' is sometimes used to indicate a 4 ft, or 48 in. (1.2 m), long tube, and '96' is used the same way for an 8 ft (2.4 m) long tube. Following the wattage number is the shape letter descriptor, usually 'T', for tubular. Following the shape descriptor is a number which is the diameter of the tube in eighths of an inch, and that completes the basic lamp designator. For example, an F32T8 lamp would be a fluorescent 32 W tube, 1 in. (2.5 cm) in diameter, understood to be standard, or 4 ft (1.2 m) long. Following the basic designator, and separated from it by a forward slash (/) are CRI and color descriptors, as well as manufacturers' specialty descriptors. For example, F32T8/841 denotes the above 32 W lamp, with a CRI of 80 or above, and 4100 K in color. High output and very high output lamps are designated with the suffix HO or VHO, respectively.

Fluorescent lamp advantages

Some of the advantages of fluorescent lighting are:

1. long life (20 000 hours average);
2. low cost;
3. high lumen to input watt ratio (F32T8/841 has efficacy of about 8 times that of standard incandescent);
4. available in a wide range of sizes and colors;
5. available with high CRI ratings;
6. low heat generation.

Fluorescent lamp drawbacks

Some of the drawbacks associated with fluorescent lighting are:

1. temperature sensitive. Output drops drastically in low temperatures. Lamps used outdoors require special low temperature ballasts;
2. require expensive dimming ballast for dimming;
3. lamps contain mercury, which is classified as an environmental hazard, and can present difficulties in disposal of burned out lamps;
4. can produce stroboscopic effect around rotating machinery, since the arc turns off and on at twice the frequency of the incoming power with magnetic ballasts.

Fluorescent lamp uses

Fluorescent lamps should be the first choice for:

1. office space ambient lighting;
2. large retail space ambient lighting;
3. interior common space lighting;
4. compact fluorescent lamps should be used in recessed downlighting and accent lighting for energy savings.

High intensity discharge (HID) lamps

High intensity discharge (HID) lamps operate on the photoluminescence principle like fluorescent lamps. That is, they require an electric arc passing through, and colliding with gas atoms to produce light. However, unlike fluorescent lamps, which produce 90% of their light from phosphors excited by the generated light, most of the light produced by HID lamps is the arc-generated light itself.

This requires that the HID lamps operate at a higher temperature, with a much higher intensity arc, and under a much higher pressure than fluorescent lamps. To do this, the HID lamp has an inner pressurized tube, called the arc tube, where the high intensity arc takes place, and an outer envelope to protect the arc tube. Like the fluorescent tube, this outer envelope can be coated with phosphors to improve the CRI of the lamp. The space between the outer envelope and the arc tube is evacuated to a high degree to maintain a constant temperature in the arc tube. High intensity discharge lamps utilize a starting electrode to ionize the gas mixture and start the arc, and a coated filament coil electrode in each end of the arc tube to maintain the arc. Like the fluorescent lamp, the HID lamp must be used in conjunction with a ballast to provide starting voltage, and to regulate the current flow once the arc has started. The ballast must be closely matched to the lamp for the system to operate properly. High intensity discharge ballasts are sensitive to momentary drops in input voltage, and for some ballasts, a drop of greater than 10% will extinguish the lamp. Figure 2.7 shows the construction of typical HID lamps.

The high intensity arc in an HID lamp will not strike until the lamp comes up to operating temperature, so there is an average warm-up time of 2–6 minutes from the time the switch is thrown until the lamp reaches full output. Conversely, if the lamp is turned off for any reason, the gasses inside the arc tube will be too hot to re-ionize immediately, and there will be a cool-down period of 5–15 minutes before the arc can be restruck. Table 2.1 is a comparison chart which shows the warm-up and restrike times for the various HID sources.

Despite these drawbacks, HID lamps are among the most efficient and long-lived lamps in use today. They are widely used to light large, high ceiling spaces, such as warehouses or gymnasiums. Included in the HID family of lamps are *mercury vapor, metal halide*, and *high pressure sodium*. A close cousin, the low pressure sodium lamp, will also be included in the HID group even though it operates at a lower arc tube pressure than the others.

Let's look now at the characteristics of each of these lamps.

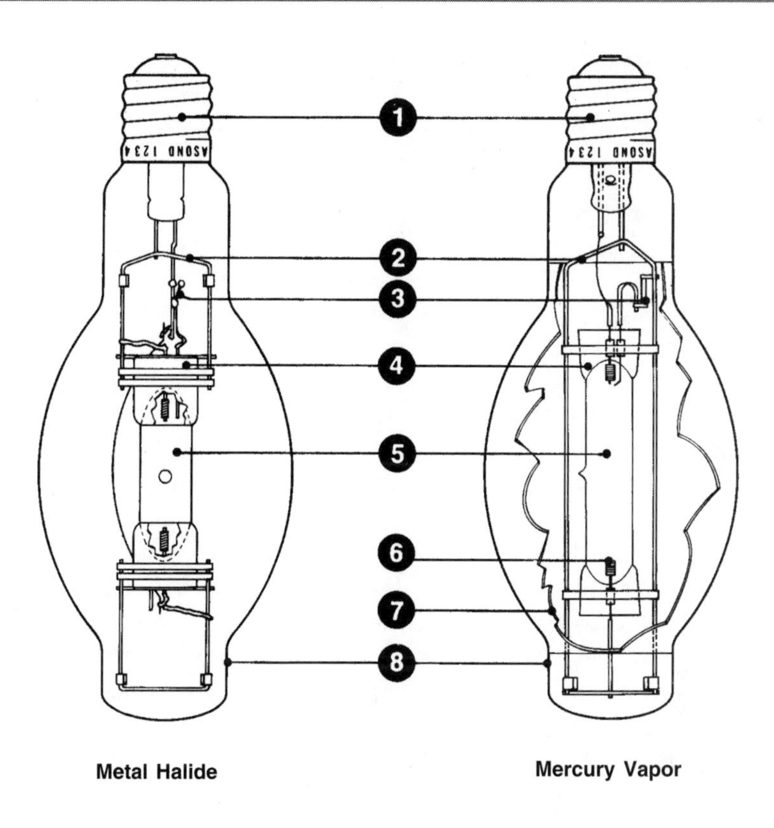

1. **Base**
2. **Supports**
3. **Starting Resistor**
4. **Arc Tube Seal**
5. **Arc Tube**
6. **Electrode**
7. **Phosphor Coating**
8. **Outer Envelope**

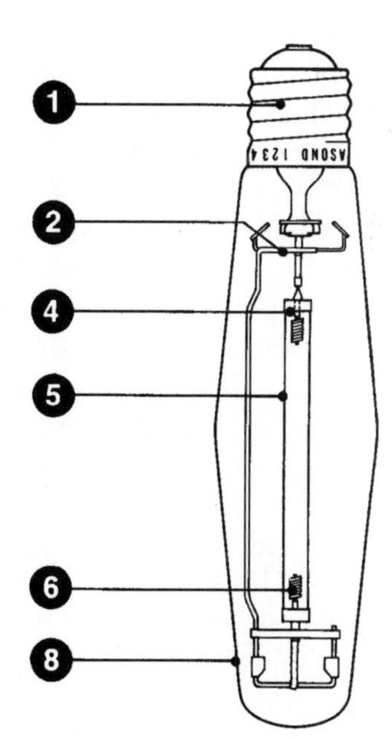

High Pressure Sodium

Figure 2.7. High intensity discharge (HID) lamp construction (source: Philips Lighting).

Table 2.1. High intensity discharge warm-up and restrike times.

Light source	*Warm-up time*	*Restrike time*
Mercury vapor	5–7 min	3–6 min
Metal halide	2–5 min	10–20 min
Pulse start metal halide	2–3 min	3–4 min
High pressure sodium	3–4 min	1–2 min
Low pressure sodium	7–10 min	3–12 s

Mercury vapor lamps

Mercury vapor lamps produce the color spectrum of mercury, which appears as a series of 'spikes' on the spectral distribution charts. The distribution for a clear mercury vapor lamp was shown in Figure 2.2.

Looking back at Figure 2.2, it can be seen that a large portion of the distribution is in the UV region, which is harmful to both the eyes and the skin. To counteract this, the outer envelope of the mercury vapor lamp is coated with a UV inhibitor. Even with the inhibitor, the clear mercury vapor lamp still produces light with a correlated color temperature from 5600 to 6400 K. In addition, there are a lot of colors that are not included in the spectrum of the lamp. This leaves the lamp with a CRI of less than 20, which offers very poor color rendering, particularly in the yellow–red region. To correct this, manufacturers have formulated phosphor coatings for the outer envelope which fluoresce in those colors that the mercury gas lacks. Two of the resulting lamps are the 'warm' mercury vapor lamp with correlated color of about 3300 K, and a 'white' lamp at about 4100 K. These lamps have a CRI of 50 or more, which is not great, but is a big color improvement over the clear lamp.

Mercury vapor lamps are the least efficient of all the HID lamps, having a lumens per input watt ratio (efficacy) lower than some of the fluorescent lamps. Mercury vapor lamps are, however, long lived, having an average life in excess of 24 000 hours. Mercury vapor lamps come in a number of shapes and sizes, including reflector lamps. The ballasts must be closely matched with the lamp in a mercury vapor system.

In the US, The American National Standards Institute (ANSI) has assigned a manufacturer's designator for mercury vapor lamps of 'H', taken from the chemical designation Hg for mercury, which (to throw in a little more useless trivia) comes from the Greek word 'hydrargyrum'. Therefore, the first letter in a mercury vapor lamp designator is H. The following numbers indicate the ballast required for use with the lamp; the two letters following the numbers indicate the physical characteristics of the lamp (i.e., size, shape, etc.). This is separated by a dash from another number, which is the wattage of the lamp. Letters following a forward slash after the wattage indicate manufacturer specialty features. Thus, an H38AV-100/DX lamp is a 100 W mercury vapor lamp requiring a type 38 ballast, having size and shape code AV and a Deluxe white phosphor coating.

Some of the beneficial features of mercury vapor lamps are:

1. long life (24 000+ hours);
2. blue–green output flattering to plant color;
3. available in a variety of sizes and shapes;
4. least expensive HID source.

Some of the drawbacks to mercury vapor lamps are:

1. long warm up time to full output (3–5 minutes);
2. poor CRI, even with phosphor coating;
3. inefficient – lowest lumen-per-watt ratio of all the HID lamps;
4. produces UV radiation which can cause skin and eye burn if the outer envelope is broken – must be shielded for indoor use.

Mercury vapor lamps are best suited for landscape or atrium lighting, where the color and long life are an asset.

Metal halide lamps

Metal halide lamps are a very efficient source of 'white' light, and are available in wide wattage range. The efficacy, or lumen output per watt input ratio for metal halide is 3–4 times that of mercury vapor, and exceeds that of fluorescent above 400 W. Metal halide also has a relatively high color rendering index, with an average CRI of about 70, and some lamps are available with a CRI above 80. Metal halide lamps come in sizes from 50 to 2000 W, with correlated color temperatures of 2900–6000 K, and in a variety of shapes.

The construction of the metal halide lamp is very similar to that of the mercury vapor lamp. The major difference between the two is the gas which fills the quartz arc tube. Where the mercury vapor lamp contains a mixture of argon gases, with mercury for the arc, metal halide lamps contain argon, mercury, and several different iodide compounds. It is the iodides which produce light of a superior spectral distribution, and give the metal halide light a high CRI. Metal halide lamps are available with a variety of phosphor coatings for the UV inhibiting outer tube to provide a desired correlated color temperature. A second type of metal halide lamp, the pulse start lamp, uses the same iodides and coatings, but instead of using a filament coil, the arc is started using a high-energy pulse generated by an igniter, similar to a high pressure sodium lamp. In addition to using an igniter circuit, the pulse start lamp differs from the standard metal halide in several other ways:

1. an improved arc tube seal allows the lamp to run hotter with higher fill gas pressure;
2. faster warm-up time (2–3 minutes);
3. faster hot re-strike time (3–4 minutes);
4. higher lumen output, and better performance (35% better than universal burn metal halide);
5. longer life than the standard universal burn metal halide lamp.

In operation, the arc in the metal halide lamp is started by either a combination of heated electrode/high starting voltages from the ballast (standard) or by a high voltage pulse from the igniter (pulsed start) to ionize the mercury in the arc tube. Once the arc starts, the iodides gradually enter the arc stream, and the output of the lamp shifts from the blue–green of mercury vapor to white. Since the arc tends to curve upward when the lamp is horizontal, some lamps are designed with a special curved arc tube to burn horizontally. Others will burn in either a vertical or horizontal (universal) position. The pulse start lamps are designed to burn in the vertical position. In addition, the pulse start lamp requires a pulse rated socket. As with the mercury vapor lamp, the ballast required for metal halide operation must be closely matched to the lamp.

The first letter in a metal halide lamp designator is 'M'. The second letter is a manufacturer's letter to denote special features of the lamp. The number following that designator is the wattage of the lamp. A forward slash usually follows the wattage, and following the slash is abbreviated data concerning the required burning position, or the base type of the lamp. A M175/U lamp then, is a 175 W metal halide lamp which will burn in universal (either horizontal or vertical) position.

Metal halide lamps offer many benefits, including:

1. highest efficacy of any 'white' light producing lamp;
2. long life (10 000–20 000 hours);
3. available in a variety of bases and shapes;
4. available in sizes from 50 to 2000 W;
5. available in a broad range of correlated color temperatures (2900–6000 K);
6. high CRI (60–90);
7. insensitive to ambient temperature.

There are also a few drawbacks associated with metal halide lamps, including:

1. requires warm-up time of 2–3 minutes;
2. must be used in a shielded fixture. Outer envelope breakage can allow emission of high levels of UV light;
3. long restrike time after power outage (5–7 minutes – standard, 2–3 minutes – pulse start);
4. relatively expensive lamp.

Metal halide is a first choice for a wide variety of applications requiring various mounting heights, including:

1. industrial facilities with high ceilings;
2. sports facilities;
3. warehouses needing high CRI;
4. retail facilities;
5. downlighting, uplighting, and accent lighting in commercial facilities;
6. ambient lighting in facilities with high ceilings;
7. outdoor building lighting.

There is a metal halide option available for almost every type of indoor luminaire. In sensitive areas, luminaires can be equipped with a second, incandescent quartz lamp, to provide illumination during the cool-down period in the event of a power dip. This is called 'quartz restrike'.

High pressure sodium lamps

High pressure sodium has the highest efficacy of any member of the true HID family, and equals or exceeds the efficacy of fluorescent in all wattages. The drawback of the high pressure sodium lamp is a low CRI, since almost 100% of the output light is in the yellow–orange region of the spectrum. Construction of the high pressure sodium lamp is similar to that of the other HID lamps, with some notable exceptions. The arc tube is made of ceramic material, rather than quartz, to withstand the corrosive effects of sodium and the extremely high temperatures required for operation. The arc tube is filled with sodium, and a small amount of mercury and xenon gas for arc starting. No starting electrodes are used in the standard sodium vapor lamp, so the ballast includes an electronic igniter circuit that works in conjunction with the transformer to provide the high voltage to start the lamp.

A hybrid sodium vapor lamp that includes the starting circuitry inside the lamp has been developed to replace mercury vapor lamps in street lighting luminaires. This lamp will start on a standard mercury vapor lamp ballast.

In operation, the lamp goes through a warm-up period of 3–4 minutes before reaching full brightness. During warm-up, the lamp undergoes several color changes as the various elements ionize. As the pressure in the arc tube increases, the lamp comes to full brightness. A power outage or voltage dip will require only approximately 1 minute for restrike.

In operation in an open luminaire, the sodium vapor lamp produces a 'narrow spotlight' light distribution pattern which focuses most of the light in a small area. A diffusion coating on the outer envelope is available for most sodium vapor lamps to spread the distribution pattern. This is the 'coated' version of the lamp. Unfortunately, this coating does not improve the CRI, which is only about 21 for most lamps, unless phosphor coatings are used.

Phosphor coatings are available for the outer envelope of lamps up to 400 W to improve the CRI to about 65, and for lamps up to 100 W to produce a CRI as high as 85.

Manufacturers have chosen to call their high pressure sodium vapor lamps by another name, ending in 'lux'. Philips, for example, calls theirs Ceramalux, while Osram–Sylvania calls theirs Lumalux. The manufacturer descriptor for Osram–Sylvania, then, would start with 'LU', for Lumalux. Following that is a number indicating wattage, and after that, a forward slash. If a 'D' follows the slash, this indicates a diffusion coating. If not, the following letters indicate base type. For example, LU50/D/MED denotes a 50 W high pressure sodium, coated, medium base lamp.

Benefits of high pressure sodium lighting are:

1. high efficacy;
2. available in sizes up to 1000 W;
3. no dangerous mercury arc in the event of outer envelope breakage;
4. normal restrike in 1 minute or less – instant restrike available;
5. insensitive to ambient temperature;
6. long life (24 000+ hours).

The main drawbacks of high pressure sodium lighting are:

1. low CRI (21 without phosphor coating);
2. relatively expensive lamp and ballast;
3. extremely high operating temperatures – must be used in fixtures which are ANSI designated for high pressure sodium use.

High pressure sodium lamps are widely used where CRI is not of great importance, but where high lumen output is. These areas include:

1. parking lots;
2. warehouses;
3. loading docks;
4. streets and highways;
5. building floodlighting.

Low pressure sodium lamps

Although not a true high intensity source, the low pressure sodium lamp is a discharge source, and is included here for completeness. Low pressure sodium is a monochromatic source, having almost 100% of its light output in two narrow bands at 589 and 589.6 nanometers in wavelength, which is in the yellow portion of the spectrum near the peak of the eye sensitivity curve. This gives low pressure sodium the lowest CRI, and the highest efficacy in the industry.

The low pressure sodium lamp is constructed much the same as the high pressure sodium lamp, except that the ceramic arc tube is bent into a 'U' shape, and dimpled to assure proper flow of the vaporized sodium.

In operation, the lamp is started on neon–argon gas, and after a warm-up period of 7–9 minutes the sodium vapor conducts, and the lamp delivers full output. If a power outage occurs, the lamp will re-strike in less than 1 minute.

The strong points of low pressure sodium are:

1. very high efficacy (as high as 200 lumens per watt);
2. low re-strike time

The drawbacks of low pressure sodium are:

1. monochromatic light output (very low CRI);
2. long warmup time;
3. relatively expensive.

Due to its low CRI, low pressure sodium has no interior applications. Its uses are primarily in large, outdoor applications where CRI is of no consequence, such as railroad yards and parking lots. Table 2.2 is a performance comparison of the four types of HID lamps.

Table 2.2. High intensity discharge (HID) performance comparison chart.

Light source	*Warm-up time*	*Restrike time*	*Efficacy (lumens/watt)*	*Color rendering*
Mercury vapor	5–7 min	3–6 min	30–55	Poor
Metal halide	2–5 min	10–20 min	60–95	Good, CRI 60–90
Pulse start metal halide	2–3 min	3–4 min	60–95	Good, CRI 60–90
High pressure sodium	3–4 min	1–2 min	60–125	Moderate
Low pressure sodium	7–10 min	3–12 s	80–180	Bad

Ballasts

Although not specifically a light source, ballasts are an integral part of the fluorescent and HID lighting systems. The ballasts used with these lamps serve two functions: (1) *they start the arc in the tube* and (2) *they regulate the current flow in the arc tube.*

Lamps and ballasts must be properly matched in order to achieve optimum (or any) operation. Let's take a look at some of the different ballasts.

Fluorescent ballasts

In general, there are three types of ballasts used with fluorescent lamps. These are: *magnetic, hybrid*, and *electronic.*

Magnetic fluorescent ballasts

The magnetic ballast is the oldest of the three types, and has been in service as long as fluorescent lamps have. The magnetic ballast is basically a transformer consisting of a laminated steel core wrapped with insulated copper or aluminum windings, and housed in a steel case which is filled with a heat dissipating 'potting' compound. In operation, the magnetic ballast supplies either line voltage to the electrode coil to start a rapid start lamp, or high voltage across the electrodes to start an instant start lamp. The windings of the transformer serve as a choke coil to limit current flow in the arc while the lamp is operating. Magnetic ballasts lower the power factor of an electrical system, and capacitors are often built into magnetic ballasts to counteract this. Magnetic ballasts containing capacitors are called *high power factor* ballasts. A magnetic ballast can control a maximum of two lamps. There is a heating, or 'ballast' loss associated with the use of magnetic ballasts. Ballast losses from magnetic ballasts can account for up to 25% of total luminaire energy consumption.

Hybrid fluorescent ballasts

Hybrid ballasts are used only for rapid start lamps. A magnetic ballast applies power to the starting coil during lamp start, and that power continues to be applied while the lamp operates. The hybrid ballast is identical in construction to the magnetic ballast, except that it contains electronic circuitry to remove the coil power during lamp operation. This reduces energy consumption by about three watts per lamp.

Electronic fluorescent ballasts

Electronic ballasts start and control current flow through fluorescent lamps through the use of electronic circuitry, rather than transformer windings. Electronic ballasts are similar to the power supplies found in computers and other electronic devices: the incoming AC power is rectified, regulated, and re-introduced into the system with different characteristics. The electronic ballast operates the lamp at a much higher frequency (20 000 Hz or greater) than the 60 Hz magnetic ballasts, and thus eliminates lamp flicker and a host of other lamp inefficiencies. Since there is no heavy steel core or windings involved in the electronic ballast, it can be made smaller and lighter than the magnetic ballast, and in operation, there is no ballast loss. One electronic ballast can control up to four lamps.

The electronic circuitry used in the ballasts tends to distort the waveform of the incoming power by injecting harmonics into the system, notably the third harmonic. To counteract this, larger power conditioning components are used in what are called *low THD* (total harmonic distortion) ballasts. A word of caution here – if the THD is *too* low (THD < 10%), the ballast inrush current goes up dramatically, and there could be switching problems with the luminaires. A ballast with a maximum THD of 15% is appropriate for most systems.

The high frequency of the electronic ballast also generates electromagnetic interference (EMI) which affects some sensitive electronic equipment such as library book detectors, power line carrier control systems, central time clock systems, and medical equipment. Hybrid ballasts should be considered for use in areas where such equipment is present. Otherwise, electronic ballasts should be the first choice for all fluorescent applications.

High intensity discharge (HID) ballasts

High intensity discharge (HID) ballasts perform the same functions that fluorescent ballasts do: they start the arc, and they regulate the arc current once the arc is started. High intensity discharge ballasts are magnetic, and each is designed to operate a particular lamp. In general, there are four types of HID ballast: *reactor, high reactance autotransformer, constant wattage*, and *constant wattage autotransformer.* Let's look at some of the features of each type.

Reactor ballast

The reactor ballast is essentially just a choke coil in series with the lamp which limits current through the lamp. The reactor ballast is used with lamps which require only line voltage to start the arc, such as mercury vapor lamps. The pros are that the reactor ballast has low inrush current, and low cost; the cons are that the ballast is very sensitive to line voltage, and lamp output can vary up to

12% with a 5% variance in voltage. With about a 10% dip in line voltage, the lamp arc will extinguish. Reactor ballasts have a low power factor, and are often used in conjunction with capacitors to remedy this. The manufacturer's symbol for a reactor ballast is 'R'.

High reactance autotransformer ballast

This ballast is an autotransformer, which means that it has two windings connected electrically, and coupled magnetically. The high reactance autotransformer ballast can boost line voltage to any starting voltage required by the lamp. The windings also serve to limit current through the lamp, as in the reactor ballast. Like the reactor ballast, the high reactance autotransformer ballast is susceptible to low line voltage, and has poor power factor. Since there is an additional winding, this ballast is more expensive that the reactor ballast. This ballast has a manufacturer's symbol of 'HX'.

Constant wattage ballast

The constant wattage ballast is a true two-winding transformer, having only magnetic coupling between the coils. This configuration is inherently insensitive to drops in the input voltage, and the ballast can sustain up to a 50% dip in line voltage before the lamp arc extinguishes. The constant wattage ballast contains a capacitor for high power factor, and the starting current does not exceed the lamp operating current. Unfortunately, these ballasts are expensive, often costing three times as much as a reactor ballast. The ballast has a manufacturer's symbol of 'CW'.

Constant wattage autotransformer ballast

This ballast is an autotransformer, with the two windings electrically connected and magnetically coupled to provide excellent voltage regulation. This ballast can withstand a 30% dip in line voltage before the lamp arc is extinguished. With its built-in capacitor for high power factor, and its low inrush current, this ballast offers the best compromise between cost and performance. Its symbol is 'CWA'.

For the more visual oriented among you, Figure 2.8 is a schematic comparison of the four types.

You should always specify constant wattage (CW or CWA) ballasts for HID luminaires which are fed from panels that also contain heavy motor loads. Industrial motors have an inrush current of 5–10 times their running current, and the inrush can last for several seconds. If the motor is large enough, this can cause a momentary voltage dip of 15–20% at the panel. Reactor ballasts connected to the panel will drop their lamps out, and the place will go dark – and remain so for the cool-down period of the lamp. It

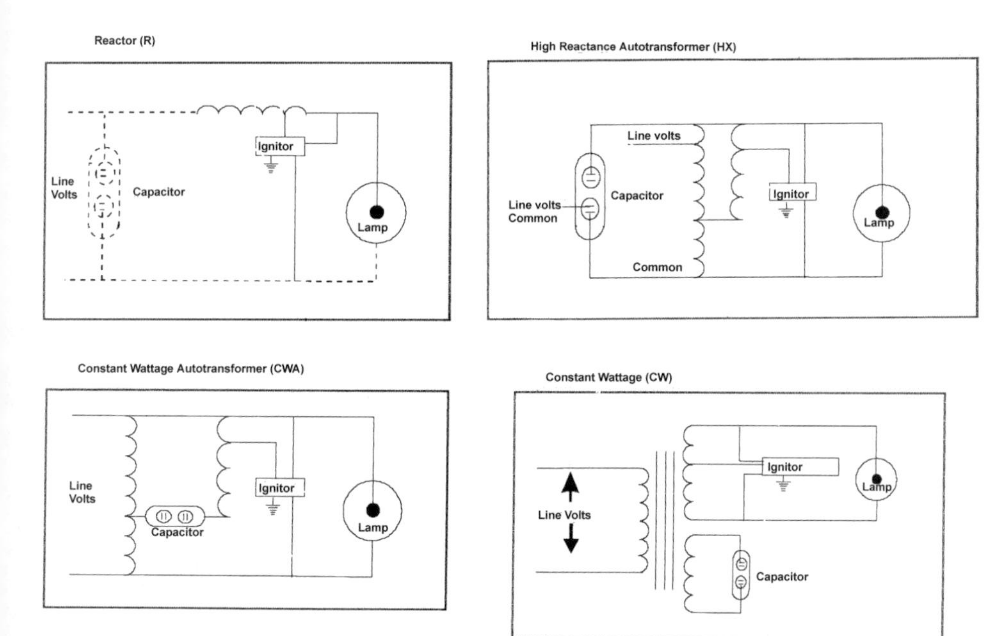

Figure 2.8. Ballast schematics (source: Advance Transformer).

is especially embarrassing to the lighting designer when the lights go out every time the compressor comes on.

Luminaires

A luminaire is a complete system for providing the kind of light we want, where we want it. The luminaire consists of the lamp, the lamp socket, the ballast, if required, the reflector, diffuser, shielding, and the housing for all those things. A luminaire needs only a mounting spot and a source of electricity to produce light.

To be an efficient producer of light, a luminaire must provide balanced, glare-free illumination. Luminaires accomplish this through the use of reflectors, which shape the light output, and shielding and diffusion devices that reduce glare and distribute the light evenly. Let's take a look at these components.

Reflectors

Reflectors are bent painted metal panels in an inexpensive fluorescent luminaire, or precisely cut or vacuum-formed mirror finish optical systems in an expensive architectural downlight, or anything in between. Reflectors serve a twofold purpose: they *shape the light pattern* of the luminaire, and they *improve the efficiency* of the luminaire (defined as the amount of light delivered by the luminaire, divided by the light produced by the lamps) by directing the usable light out of the luminaire onto the work surface. Glossy or mirrored reflectors produce direct light, while matte finish reflectors produce scattered, or diffuse light.

Factors affecting the performance of a reflector are *the reflectivity of the reflector material* and *the optical geometry of the reflector*. Reflectivities range from about 60% for standard white paint to 95% for polished aluminum. The geometry of the reflector largely determines the shape of the light pattern emitted by the luminaire.

Over the years, luminaire manufacturers have experimented with various shapes of reflectors to improve performance. One shape in particular which has proven effective in ceiling-mounted fluorescent fixtures is the parabola: a reflective trough in the shape of a parabola is installed lengthwise in the luminaire, and the fluorescent tube is mounted at the apogee of the parabola. Figure 2.9 is a ray trace of the parabolic reflector.

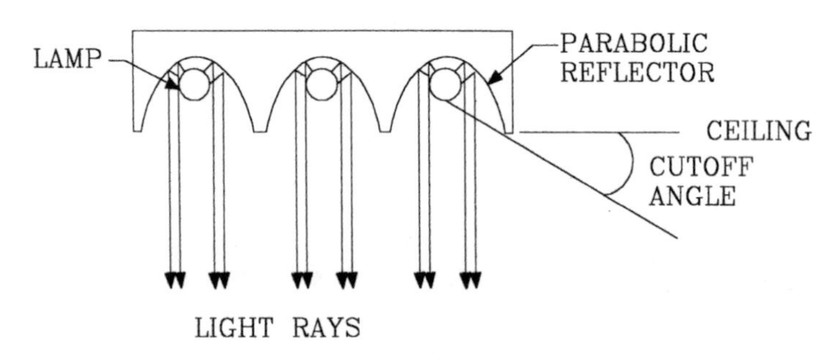

Figure 2.9. Parabolic reflector ray trace.

Another interesting geometric shape used for lighting purposes is the ellipse. When used in a recessed downlight, an elliptical reflector can be designed such that its focal point is exactly at ceiling level. The focal point in an elliptical reflector is that point at which all reflected light rays converge. Past that point, they diverge again. This allows all the light produced by the lamp to be passed through a very small opening in the ceiling. This is called a 'pin-hole' luminaire, and its ray trace is shown in Figure 2.10.

Shielding and diffusion devices

Shielding devices are used to reduce the glare produced by a luminaire; diffusion devices are used to diffuse the produced light, and to conceal the bare lamps of the luminaire.

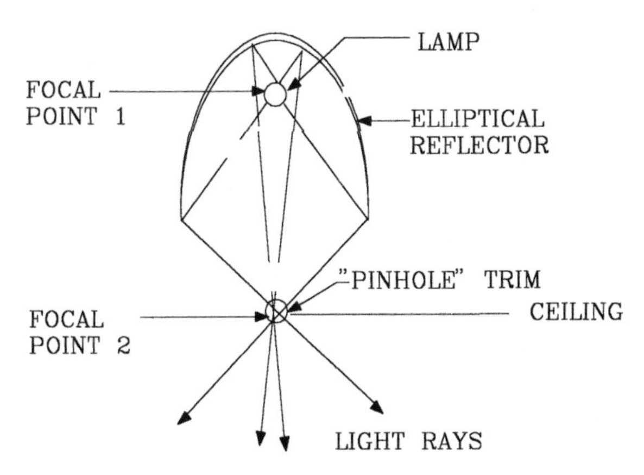

Figure 2.10. Elliptical reflector ray trace.

Baffles

An example of a shielding device is the baffles used in conjunction with the parabolic reflectors in a recessed parabolic luminaire. Figure 2.11 illustrates this luminaire.

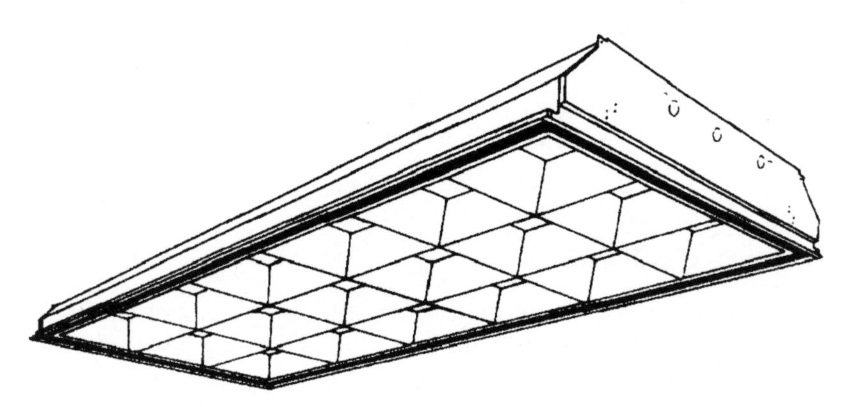

Figure 2.11. Baffles in a parabolic luminaire (source: Lithonia Lighting).

It is seen that the baffles running along and perpendicular to the sides of the parabolic trough form multiple 'cells' along the length of the luminaire. These cells provide sharp cut-off glare shielding when the luminaire is viewed at an angle. The smaller the cells are, the better the control. However, since the baffles are made of opaque material, small cells tend to mask much of the light produced by the luminaire, and thereby reduce its efficiency.

Diffusers

Diffusers used with fluorescent fixtures also influence the distribution pattern of the luminaire. The most common diffuser employed in fluorescent fixtures used for general area lighting is a flat plastic sheet embossed with some sort of diffusion pattern. The two most common plastics used for this purpose are *polycarbonate* and *acrylic*. Polycarbonate is exceptionally strong, but unless specially treated to resist the UV emissions from the lamp, will turn yellow after a period of time. Acrylic, though not as strong, is impervious to UV. Acrylic diffusers are made as thin as 0.10 in. (0.25 cm) thick for a 2 ft × 4 ft (0.6 m × 1.2 m) unit, but since this material is not especially strong, sheets this thin will tend to sag in the luminaire. It is always best to specify UV-treated polycarbonate, or acrylic diffusers no less than 0.125 in. (0.318 cm) thick. There is a variety of diffuser styles available in either material which can produce a desired distribution pattern.

Luminaire housings

Housings for luminaires can be metallic, such as sheet steel or aluminum for general purpose, or plastic for corrosive or high abuse areas. Luminaires that fit into standard 2 ft × 4 ft grid (0.6 m × 1.2 m) ceilings are available with air-handling housings, which can serve as return or supply registers for the building HVAC (heating, ventilating, and air conditioning system). A benefit of this type of housing is that luminaire heat is removed before entering the space.

Luminaire classifications

Luminaires used for area lighting have been classified by the Illuminating Engineering Society (IES) according to their distribution pattern and by the downward and upward component of the light generated. In general, there are five categories of luminaire: *direct, semi-direct, general diffuse, semi indirect,* and *indirect*. The distinguishing patterns of four of these types is shown in Figure 2.12a–d.

The general diffuse lighting pattern is one that contains equal components in all directions.

Direct illumination

Direct illumination, as shown in Figure 2.12a, is the most efficient use of the luminaire to provide lumens to the work area. It also produces the greatest amount of glare, unless shielded. Direct illumination is best used where you desire a high ambient lighting level, such as kitchens, day care centers, and general work spaces. Direct

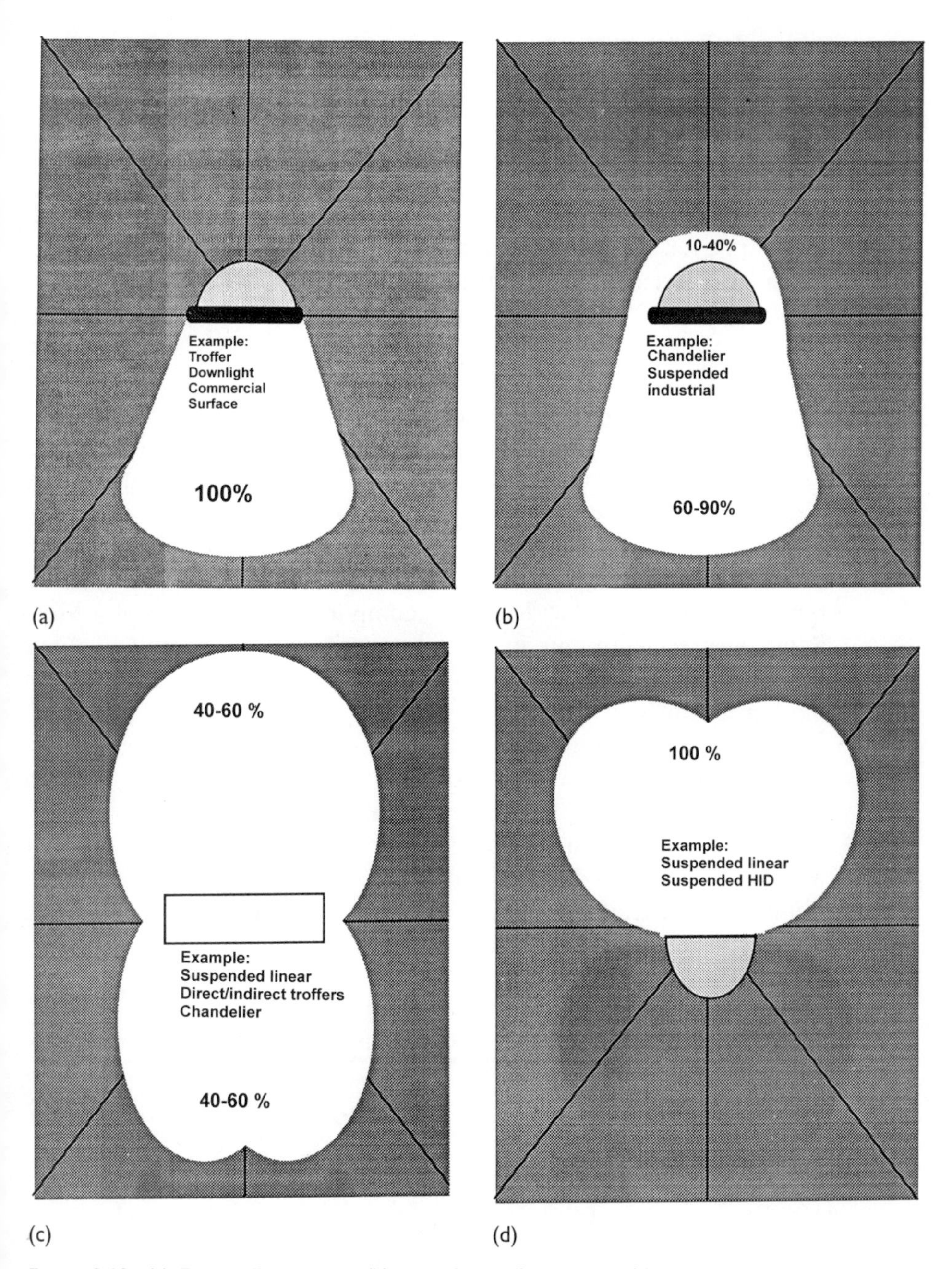

Figure 2.12. (a) Direct illumination, (b) semi-direct illumination, (c) semi-indirect illumination, (d) indirect illumination (source: F.H. Jones).

illumination is also used for task lighting, where a high lighting level in a limited area is desired. When used with parabolic reflectors and baffles, or cutoff shielding, direct illumination is acceptable for use in areas where glare is undesirable, such as offices containing video data terminals. Direct luminaires may be recessed into the ceiling, surface-mounted on it, or suspended below it.

Semi-direct illumination

Semi-direct luminaires have an upward component, as shown in Figure 2.12b, so they must be either surface-mounted or suspended. They are used for the same purpose as direct luminaires, except that when suspended, the upward component lights the ceiling, and eliminates the 'cave effect' produced by the suspended direct luminaire.

Semi-indirect and indirect illumination

The semi-indirect and indirect luminaires must be suspended to utilize the upward reflected component of light. These luminaires differ only in the percentages of uplight and downlight that they produce. Indirect luminaires produce almost 100% uplight, which provides a diffuse, glare-free illumination particularly well suited for use in a room with heavy computer terminal usage. Most manufacturers offer an indirect luminaire which produces about 10% direct illumination to eliminate the dark area created by the luminaire housing.

Wall-mounted luminaires are classified in the same manner as ceiling-mounted ones, and have basically the same illumination patterns. One type of wall-mounted luminaire that receives considerable use in modern buildings is the cove light, which is usually mounted near the ceiling, and is used in an indirect pattern to illuminate the ceiling. Figure 2.13 shows a typical cove lighting installation.

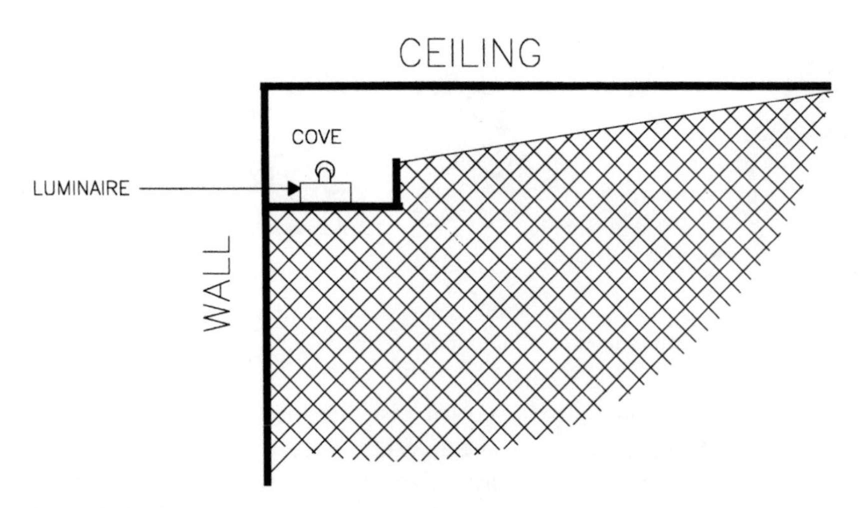

Figure 2.13. Cove lighting (source: author).

Luminaire photometric data

Manufacturer's published information for their luminaires includes a graphic representation, in two axes, of the distribution pattern produced by each luminaire. This representation is called a *candle-power distribution curve* for the luminaire. It is also called a 'batwing' curve in the industry, because of the distinctive shape of some curves. Figure 2.14a and b shows candlepower distribution curves for a recessed HID direct pattern downlight luminaire, and a recessed fluorescent direct pattern downlight luminaire, respectively.

These curves were generated in test laboratories by taking candle-power measurements around a test luminaire and plotting them on polar coordinates. In applications where all the light from the luminaire is projected in one direction, such as the direct distribution of the luminaires in Figure 2.13, only half of the polar graph is shown. We can use this published data to give us some idea of the light pattern which a particular luminaire will provide. Distribution curves are a valuable tool for quickly evaluating luminaire suitability, especially for task lighting applications.

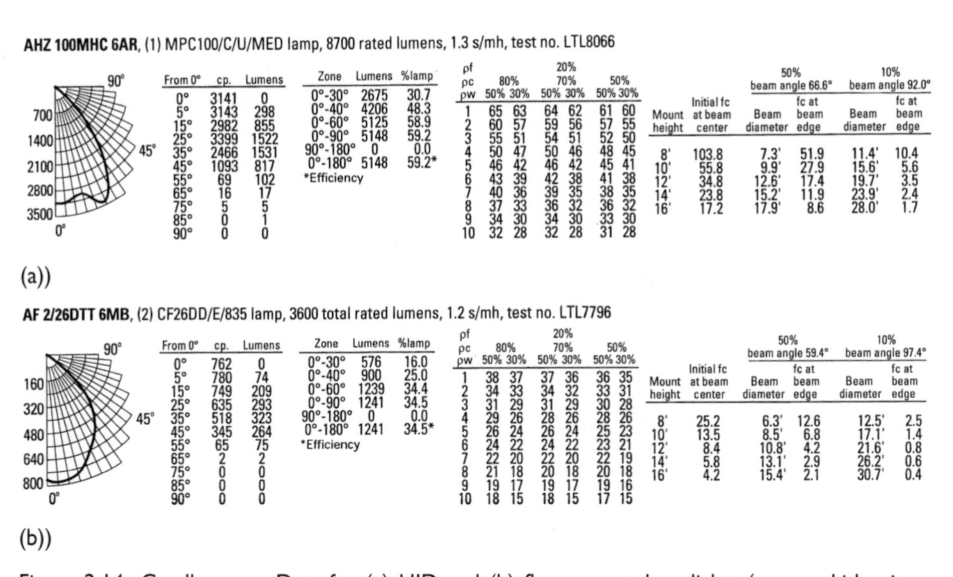

AHZ 100MHC 6AR, (1) MPC100/C/U/MED lamp, 8700 rated lumens, 1.3 s/mh, test no. LTL8066

From 0°	cp.	Lumens
0°	3141	0
5°	3143	298
15°	2982	855
25°	3399	1522
35°	2466	1531
45°	1093	817
55°	69	102
65°	16	17
75°	5	5
85°	0	1
90°	0	0

Zone	Lumens	%lamp
0°-30°	2675	30.7
0°-40°	4206	48.3
0°-60°	5125	58.9
0°-90°	5148	59.2
90°-180°	0	0.0
0°-180°	5148	59.2*

*Efficiency

ρf	20%					
ρc	80%		70%		50%	
ρw	50%	30%	50%	30%	50%	30%
1	65	63	64	62	61	60
2	60	57	59	56	57	55
3	55	51	54	51	52	50
4	50	47	50	46	48	45
5	46	42	46	42	45	41
6	43	39	42	38	41	38
7	40	36	39	35	38	35
8	37	33	36	32	36	32
9	34	30	34	30	33	30
10	32	28	32	28	31	28

		50% beam angle 66.6°		10% beam angle 92.0°	
Mount height	Initial fc at beam center	Beam diameter	fc at beam edge	Beam diameter	fc at beam edge
8'	103.8	7.3'	51.9	11.4'	10.4
10'	55.8	9.9'	27.9	15.6'	5.6
12'	34.8	12.6'	17.4	19.7'	3.5
14'	23.8	15.2'	11.9	23.9'	2.4
16'	17.2	17.9'	8.6	28.0'	1.7

(a))

AF 2/26DTT 6MB, (2) CF26DD/E/835 lamp, 3600 total rated lumens, 1.2 s/mh, test no. LTL7796

From 0°	cp.	Lumens
0°	762	0
5°	780	74
15°	749	209
25°	635	293
35°	518	323
45°	345	264
55°	65	75
65°	2	2
75°	0	0
85°	0	0
90°	0	0

Zone	Lumens	%lamp
0°-30°	576	16.0
0°-40°	900	25.0
0°-60°	1239	34.4
0°-90°	1241	34.5
90°-180°	0	0.0
0°-180°	1241	34.5*

*Efficiency

ρf	20%					
ρc	80%		70%		50%	
ρw	50%	30%	50%	30%	50%	30%
1	38	37	37	36	36	35
2	34	33	34	32	33	31
3	31	29	31	29	30	28
4	29	26	28	26	28	26
5	26	24	26	24	25	23
6	24	22	24	22	23	21
7	22	20	22	20	22	19
8	21	18	20	18	20	18
9	19	17	19	17	19	16
10	18	15	18	15	17	15

		50% beam angle 59.4°		10% beam angle 97.4°	
Mount height	Initial fc at beam center	Beam diameter	fc at beam edge	Beam diameter	fc at beam edge
8'	25.2	6.3'	12.6	12.5'	2.5
10'	13.5	8.5'	6.8	17.1'	1.4
12'	8.4	10.8'	4.2	21.6'	0.8
14'	5.8	13.1'	2.9	26.2'	0.6
16'	4.2	15.4'	2.1	30.7'	0.4

(b))

Figure 2.14. Candlepower Data for (a) HID and (b) fluorescent downlights (source: Lithonia Lighting).

Lighting calculations

The distance from the center point of the candlepower distribution curve to any point on the curve gives the lumens of the luminaire in that direction. For asymmetric luminaires, such as a 2 ft × 4 ft (0.6 m × 1.2 m) fluorescent luminaire, data is given for each of the two axes of the fixture: along the length of the luminaire, or parallel to it; and across the luminaire, or perpendicular to it.

Now, if you remember the inverse square law and the cosine law of incidence from Chapter 1, and you know the distance and direction of your chosen luminaire from a surface that you want to examine, you can calculate the footcandles on that surface. For example, suppose you were interested in finding the footcandles produced on a desk by the fluorescent luminaire in Figure 2.14b. The center of the desk is 10 ft from the luminaire, and at an angle of 30°. You can look at the distribution curve and read the lumens at 30°. Using the cosine law of incidence, $I = L \cos X/D^2$, the illumination produced at the center of the desk by this one luminaire is: Lumens × 0.866/100, for footcandles on the square foot at the center of the desk.

Some manufacturers produce the lumen values at the various angles in tabular form, and the total lumens in an angular zone are shown. The angles listed are measured from the vertical, or normal to the luminaire. Figure 2.15 is one such set of distribution data.

As seen in Figure 2.15, a lot of other luminaire information is included in the manufacturer's data, or 'cut' sheets. The coefficient of utilization (CU) of the luminaire, as shown in the table, is a ratio of the lumens which reach an assigned working plane to the total lumens generated by the lamps in the luminaire. Many factors contribute to the coefficient of utilization, including luminaire efficiency and distribution pattern, room geometry and reflectances, and luminaire mounting height. The coefficient of utilization is useful for general area calculations, such as the lumen, or zonal cavity calculation method, which yields an approximation of the average lumens per square foot of the space under examination.

Zonal cavity calculations

Zonal cavity calculations are a quick manual method of estimating average footcandles, or the number of luminaires required to produce a desired average footcandle level within a defined space. To get some idea of how a zonal cavity calculation works, let's take a hypothetical office space which is 20 ft × 20 ft, with a 9 ft ceiling. The interior designer has selected a white ceiling tile with a reflectance of about 0.8, a light gray semi-gloss wall paint with a reflectance of about 0.7, and a darker gray carpet with a reflectance of 0.2. The work plane is the desktop, at 2.5 ft above the floor.

FEATURES

- Door is fully gasketed flush steel with mitered appearance. Corners screwed together for rigidity, easy lens maintenance.
- Optional aluminum door frames available.
- Fixture sides feature rolled edge for safer handling.
- T-hinges die-formed for maximum strength. Hinges or latches from either side.
- Opposing, rotary-action cam latches standard for secure door closing. Latches painted after fabrication for smooth finish.
- Integral T-bar safety clips hold T-bar securely, no fasteners required to install.
- Snap-in socket tracks allow socket replacement without tools.
- Full-depth end plates embossed for rigidity.
- Urethane foam gasket eliminates light leaks between door frame and housing.
- Guaranteed for one year against mechanical defects in manufacture.

SPECIFICATIONS

BALLAST — Thermally protected, resetting, Class P, HPF, non-PCB, UL listed, CSA-certified ballast is standard. Energy saving and electronic ballasts are sound rated A. Standard combinations are CBM approved and conform to UL 935.

WIRING & ELECTRICAL — Fixture conforms to UL 1570 and is suitable for damp locations. AWM, TFN or THHN wire used throughout, rated for required temperatures.

MATERIALS — Housing formed from cold-rolled steel. Acrylic shielding material 100% UV stabilized. No asbestos is used in this product.

FINISH — Five-stage iron-phosphate pretreatment ensures superior paint adhesion and rust resistance. Painted parts finished with high-gloss, baked white enamel.

LISTING — UL listed and labeled. Listed and labeled to comply with Canadian and Mexican Standards (see options).

Specifications subject to change without notice.

Catalog Number	Type

Static Grid Troffer

GT 2'x4'

2, 3 or 4 lamps

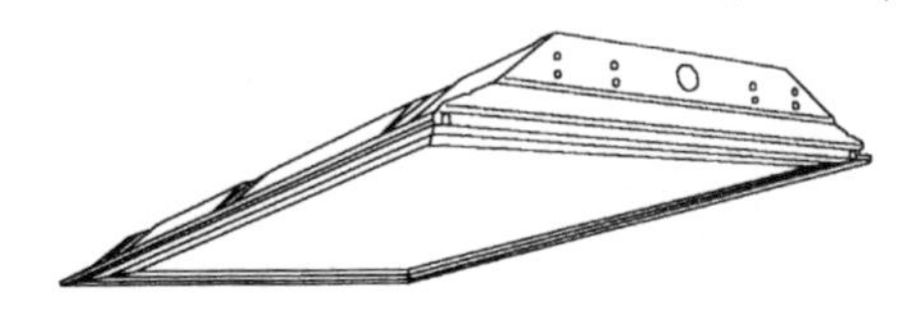

PHOTOMETRICS

Calculated using the zonal cavity method in accordance with IESNA LM41 procedure. Floor reflectances are 20%. Lamp configurations shown are typical. Full photometric data on these and other configurations available upon request. LER (Luminaire Efficiency Rating) calculated in accordance with NEMA standard LE-5. See sheet number LER for details.

2GT 3 32 A12
Report LTL 5546
S/MH (along) 1.2 (across) 1.3
Coefficient of Utilization

Ceiling	80%			70%			50%		
Wall	70%	50%	30%	70%	50%	30%	50%	30%	10%
1	82	79	76	80	77	75	74	72	70
2	76	70	66	74	69	65	66	63	60
3	70	63	58	68	62	57	60	55	52
4	65	56	51	63	56	50	54	49	45
5	59	50	44	58	50	44	48	43	39
10	40	30	24	39	30	24	29	24	20

Zonal Lumens Summary

Zone	Lumens	%Lamp	%Fixture
0-30	2019	23.2	31.1
0-40	3320	38.3	51.1
0-60	5495	63.2	84.7
0-90	6491	74.6	100.0
90-180	0	0.0	0.0
0-180	6491	74.6	100.0

LER = 62

2GT 4 32 A12
Report LTL 4866
S/MH (along) 1.2 (across) 1.3
Coefficient of Utilization

Ceiling	80%			70%			50%		
Wall	70%	50%	30%	70%	50%	30%	50%	30%	10%
1	84	81	78	82	79	76	76	74	72
2	77	72	67	75	70	66	68	64	61
3	71	64	59	70	63	58	61	57	53
4	66	58	52	64	57	51	55	50	46
5	60	51	45	59	51	45	49	44	40
10	41	31	25	40	30	25	30	24	21

Zonal Lumens Summary

Zone	Lumens	%Lamp	%Fixture
0-30	2757	23.8	31.3
0-40	4532	39.1	51.4
0-60	7476	64.5	84.9
0-90	8809	75.9	100.0
90-180	0	0.0	0.0
0-180	8809	75.9	100.0

LER = 62

2GT 4 40 A12
Report LTL 4432
S/MH (along) 1.2 (across) 1.3
Coefficient of Utilization

Ceiling	80%			70%			50%		
Wall	70%	50%	30%	70%	50%	30%	50%	30%	10%
1	76	73	70	74	71	69	69	67	65
2	70	65	61	68	64	60	61	58	56
3	65	58	53	63	57	53	55	51	48
4	60	52	47	58	51	46	50	45	42
5	55	47	41	54	46	41	45	40	36
10	37	28	23	36	28	22	27	22	19

Zonal Lumens Summary

Zone	Lumens	%Lamp	%Fixture
0-30	2739	21.7	31.6
0-40	4494	35.7	51.8
0-60	7372	58.5	85.0
0-90	8675	68.8	100.0
90-180	0	0.0	0.0
0-180	8675	68.8	100.0

LER = 62

Figure 2.15. Candlepower data, Lithonia 2GT432 (source: Lithonia Lighting).

The luminaire that we will use for the office is the 2GT432A12 fluorescent luminaire of Figure 2.15. The luminaire will be mounted in the ceiling grid.

The zonal cavity method entails dividing the space into three cavities: the ceiling cavity, which is that space above the luminaire; the room cavity, which is the space between the luminaire and the work plane; and the floor cavity, which is the space below the work plane. We have no ceiling cavity in this example, since the luminaires are mounted directly in the ceiling. The room cavity extends from the desk top to the ceiling, a distance of 9 ft - 2.5 ft = 6.5 ft. The floor cavity is 2.5 ft deep. We are primarily interested in illuminating the room cavity, since that's where all the activity will take place. Let's hypothesize a little further and say that we have defined the lighting criteria for this office space to be 70 footcandles (70 fc) at the desktop. (You'll learn more about defining lighting criteria in the next chapter.) What we need to find out by our zonal cavity calculation is how many luminaires will be required in this space to provide that 70 fc.

The zonal cavity method allows us to calculate a cavity ratio for the cavity under examination, which in our case is the room cavity. By calculating the room cavity ratio (RCR) for our space, we will be able to select the proper coefficient of utilization using the manufacturer's tables. The formula for calculating a cavity ratio is:

$$\frac{CR = 5H \times (\text{room length} + \text{room width})}{(\text{room length} \times \text{room width})}$$

where H is cavity height.

In our case, the RCR of the office is:

$$\frac{5(6.5\text{ ft}) \times (20\text{ ft} + 20\text{ ft})}{(20\text{ ft} \times 20\text{ ft})} = 3.5$$

Now, if we go back to the coefficient of utilization table in Figure 2.15, we see that, with a RCR of 3.5, the CU can be interpolated as 0.645, using the reflectances mentioned earlier.

The zonal cavity method calculates average illumination by determining the total number of usable lumens produced by the lighting system, and dividing that number by the number of square feet in the space. This yields an average lumen/sq. ft (sq. m), or footcandle value at the work surface. Conversely, if the desired footcandles is given, the number of luminaires required to produce those footcandles may be calculated.

The usable lumens produced by a luminaire in a space depends upon a number of things: first is the maximum number of lumens that the luminaire is capable of producing. This depends on how many lamps the luminaire has, how many lumens each lamp will produce, and the ballast factor (BF) of the ballast. Manufacturer's data for the luminaire indicates the number of lamps in the luminaire, and often lists lamp lumens, as well. If not, lumens can be found in lamp catalogs.

Easy for me to say, right? OK – you won't find a lamp catalog in the public library, maybe not even in a college library, but this is the twenty-first century. We've got the Internet. Most lamp manufacturers have a Web site, and we'll find out how to access them a little later.

Ballast factor is a decimal number, usually less than 1, which compares the performance of the ballast in the luminaire with that of the standard laboratory ballast used to determine the lamp lumen output listed in the lamp catalogs. Ballast factor can be specified, or obtained from the luminaire manufacturer. A BF of less than 1 means that the lamps will produce less than the listed lumen output, since actual output is obtained by multiplying listed output times the BF. Ballasts are available with BF of greater than 1, which means that the lamps will produce *more* than listed output. In general, the lower the BF, the less the lumen output, and the less energy the luminaire will use. If the BF cannot be obtained from the manufacturer, use 0.80.

Using all of the above factors, we can design for an approximate *initial* lighting level in the space. If we want to design for a *maintained* lighting level, we must consider some things that will happen after the system is installed. The first of these is lamp lumen depreciation (LLD). As a lamp ages, its light output becomes less. The major lamp manufacturers have done sufficient testing to be able to predict fairly accurately how their lamps will perform over time. The normal design point is at 70% of the lamp's lifetime, and the manufacturers usually list a *mean*, or *design* lumen value for their lamps at that point. A typical 32 W T8 fluorescent lamp which operates 12 hours per start has an LLD of 0.86. Another factor which affects fixture performance over time, but which is harder to quantify, is lumen dirt depreciation (LDD) of the fixtures and the surfaces in the space. No matter how many charts and graphs you may see on this subject, LDD is dependent on the use of the space, and the quality of the maintenance of that space. In our office space, if the maintenance staff changes the HVAC filters regularly, and if they clean the luminaire each time they change a lamp, we could expect an LDD of about 0.85.

Now that we've thought through how our system will perform in the office space, let's use the zonal cavity method to calculate the number of luminaires of the type shown in Figure 2.15 that will be required to give us an average of 70 fc throughout the space.

We have said that the average FC level =

$$\frac{\text{Total usable lamp output (lumens)}}{\text{Area of the space (sq. ft)}}$$

And the usable lamp output is:

(No. of Luminaires) × (Lamps/Luminaire) × (Lumens/Lamp) × BF × CU × LLD × LDD

So, by doing a little mathematical gymnastics, we can say that

$$\text{(No. of Luminaires)} = \frac{\text{Desired Average FC Level} \times \text{Area}}{\text{(Lamps/Luminaire)} \times \text{(Lumens/Lamp)} \times \text{BF} \times \text{CU} \times \text{LLD} \times \text{LDD}}$$

All we have to do now is plug in the values, and turn the crank:

Desired footcandles = 70
Area = 20 ft × 20 ft = 400 sq ft
Lamps/Luminaire = 4
Lumens/Lamp = 2850 lumens – from lamp catalog data (we will see later how to get this).
BF = 0.80 (assumed BF for generic electronic ballast)
CU = 0.68
LDD = 0.80
LLD = 0.86

Plugging all this in, we get:

$$\text{No. of luminaires} = \frac{70 \times 400}{4 \times 2850 \times 0.8 \times 0.68 \times 0.80 \times 0.86} = 6.56$$

Since we cannot buy 0.56 of a luminaire, we'll use six luminaires and get a little less than 70 fc.

Now that we've done all this high level calculating, with decimal points and everything, we can be sure that, if we put six luminaires in the ceiling of that space, we'll have 70 fc at 30 in. above the floor everywhere in the room, right? *Wrong*! If you took a light meter into the space after the six luminaires were installed, you would be hard pressed to find a 70 fc reading anywhere in the room. That would only happen if the output from those six luminaires were spread evenly throughout the ceiling of the room, because what we have calculated is *average* illumination. As it is, we have 48 square feet of light source (six individual 2 ft × 4 ft luminaires) to place in 400 square feet of ceiling. If we put them all in one corner of the room, we would have considerably more than 70 fc there, but much less everywhere else. We would still have an *average* of 70 fc in the room.

What we try to do in general area illumination situations like this is to distribute the six luminaires such that their spacing from each other, and from the walls is approximately equal, to get the best distribution of light possible. Since the luminaires are asymmetric, we measure our spacing from the center point of the luminaire.

The manufacturer's information for our luminaire, shown in Figure 2.15, indicates the spacing criterion, S/MH, or maximum distance between luminaires, is 1.2 × mounting height in the long direction. Note that this is the mounting height above the working plane, not the floor. In our case, this is 1.2 × 6.5 ft, or approximately 8 ft. It is always best to maintain the published spacing criterion in your designs, in order to achieve relatively even illumination throughout the space.

Now let's lay out those six luminaires in the space, and calculate the illumination at 2.5 ft from the floor on a 1 ft × 1 ft grid everywhere in the space, using the cosine law of incidence. Bear in mind, that there are six luminaires, so we have to calculate the contribution of each luminaire at each point on the grid, and then add them up to get the total illumination at that point. This is a nice little exercise to do on a two-week holiday, *or* we can call in the computer.

Computerized calculations

There are a number of commercially available point-by-point lighting calculation programs, and many of them are based around the ray tracing calculation engine of the LUMEN MICRO program, the original calculation program developed by Lighting Technologies, Inc. Others are based around a luminous rate of transfer calculation engine, such as the one that we will use in our calculations. For our computer calculations throughout this book, we will use VISUAL BASIC, a program developed by Lithonia Lighting, a large lighting company located in Conyers, GA, USA. VISUAL BASIC is available as a free download on Lithonia's Web site, www.lithonia.com. We should now get on the Internet, and go to this site.

From the opening welcome screen, we select the VISUAL download section, and in this section, we see the software options. Several offerings appear here, the 'Visual 2.0 SP1' option, the 'Photometrics' option, and the 'Basic edition users guide' option. The Visual 2.0 SP1 package contains the calculation package, and we should download the package that is compatible with our computer's operating system. Once this download is complete, we need to download the users guide for reference. This will give us a calculation program, and instructions for using it. The next thing we need to do is to download some photometrics for luminaires, which we will use in our exercises. We go back to the 'Photometrics option', and go into the Lithonia library of IES files. These files provide the performance data for the luminaires, which we will be using in our examples. This data is in digital IES format, and it is the same data that is published in the manufacturer's performance charts, such as that shown in Figure 2.15.

In the Lithonia library, we see that data is available for many different types of lighting. You can download the whole library if you have a lot of free hard drive space, and you will be able to run calculations using any luminaire made by the Lithonia group. For our immediate purposes, though, we only need to download the fluorescent lighting group, which contains architectural downlights, lensed troffers, parabolics, etc.

Once you complete these downloads, you need to install the VISUAL BASIC program by clicking on the self-extracting '.exe' file which was downloaded. Now we are ready to run a computer calculation. (Note: Web sites are changed from time to time, so the procedure above may not exactly fit the Web site as it exists when you visit it. All of the information will be there, though, and a little navigating around the site should allow you to download the latest versions of the files.)

The VISUAL program was developed by the Lithonia Lighting group as a design tool for lighting designers and specifiers. It is intended for use on personal computers, utilizing catalog and photometric data for all the luminaires manufactured by the Lithonia group, both interior and exterior. It will also use photometric data from other manufacturers, provided that the data is in IES format. The design program contains calculation engines and layout modeling capabilities.

There is an advanced version of the VISUAL program, called VISUAL PRO, which contains many more features than the VISUAL BASIC program, including the ability to import digital drawing files to allow lighting layout and calculations to be performed using actual floor plans, or site plans. VISUAL PRO is available for purchase from Lithonia Lighting.

For our example office space, we will use the internal layout editor in our downloaded VISUAL BASIC program. Just fill in the blanks in the layout screens to input the room geometry and reflectances. For a ceiling type, we'll use a 2 ft × 2 ft grid.

The program allows the designer to select photometric data from any of the luminaires in the Lithonia group to use for the calculations by entering an electronic 'catalog' and selecting the desired luminaire. You do this by selecting the file icon next to the 'photometric file' input field. Then select 'fluorescent', 'lensed troffers', and you'll see all the lensed troffers that Lithonia makes. Select a 2GT432 troffer as seen in Figure 2.15, and the program will load the same photometric data that is represented in tabular form in Figure 2.15, except that it is in IES standard format electronic form. We select the standard LLF for a lensed troffer, and put in the desired 70 fc. Our layout now shows six luminaires, which is about what we would expect, since our zonal cavity calculation showed that we need 6.3 luminaires. We can now run the calculator and find out the point-by-point footcandle values for each point on our 1 ft × 1 ft grid in a flash. How clever we are. Wait a minute... the average footcandle reading for this arrangement is higher than the zonal cavity calculation! How could this be? Well, this is because the particular luminaire that we selected has a little better performance than the generic one we conjured up for the zonal cavity calculations. This is another benefit of computer calculations. You can use the real IES performance data for the real luminaire instead of interpolating from a generic photometric chart. The results of our computer run are shown in Figure 2.16.

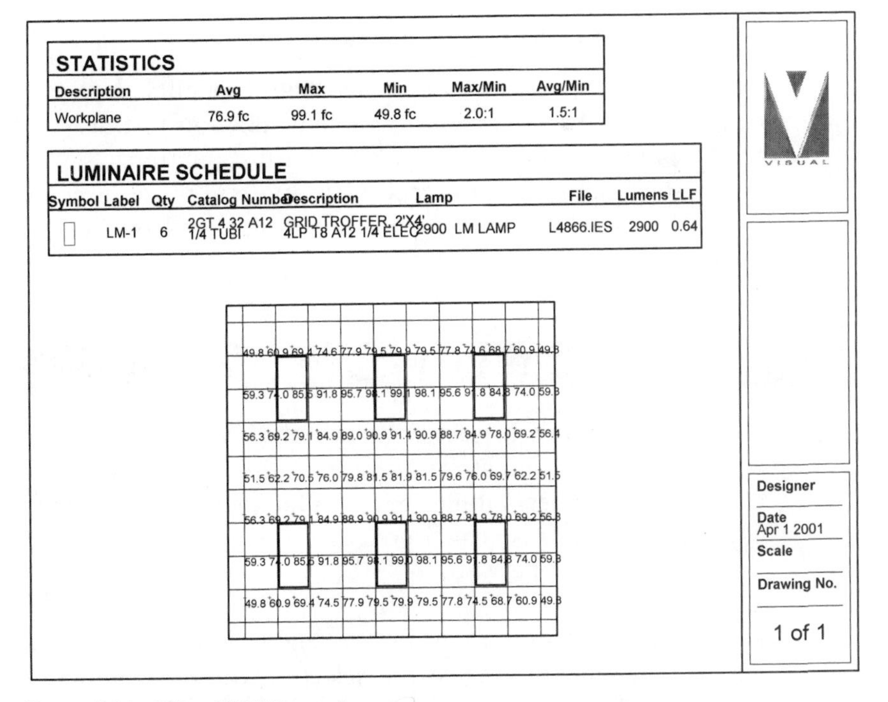

Figure 2.16. Office VISUAL results.

Clearly, the computer is the way to go if you are looking for accurate illumination values for a space, especially if you're using more than one type of luminaire, or if there are obstacles in the space. Still, the zonal cavity calculation method can provide a quick average illumination estimate if you don't have your computer handy.

Controls

Controls are what you use to make your lighting system behave as you want it to. Controls are discussed in depth in Chapter 4, but it will be useful to know the basic concept behind the devices, so you can do the exercises in the next chapter. The most commonly used forms of lighting control are:

- Switches – on–off manual devices that control the power to the system.
- Dimmers – devices that reduce the light output of the luminaires in the system; incandescent lamps are dimmed directly, fluorescent lamps require a special dimming ballast.
- Photocells – devices that turn the luminaires on when no light is present, and off when light is present.
- Timers – devices that turn the lights on and off at preset times.
- Occupancy sensors – devices that turn the lights on when the space is occupied, and off when it isn't.

And that about covers it. A quick bagful of nifty devices that you can use to make your lighting system function just as you want it to. You're only limited by budget and imagination.

Now let's look at another kind of limits – those imposed by the regulatory agencies that impact our design and the way that it is built.

Standards, codes, and design guidelines

(Note: The following section lists only US Codes and Standards, and is included to illustrate the application of standards in lighting design. The reader should become familiar with the comparable codes and standards which govern in his or her locale, and apply them in a similar fashion)

Although not design tools *per se*, lighting standards, codes, and design guidelines influence the selection of lamps and luminaires, and can set the tone for the design. It is useful to the designer to be familiar with the standards and codes with which the design must comply. It is also useful to be aware of recognized design guidelines, and to take advantage of the many hours spent by others in developing them. Let's look now at some of these standards and guidelines that apply directly to interior lighting design.

NFPA-70: The National Electric Code

The National Electric Code (NEC) is the governing code for all the electrical installations within the US. Virtually every county in every state has adopted the NEC in whole, or in part, into their own electrical code. You can be sure, as a designer, that the installation of your design will come under the scrutiny of the building inspector – who is looking for violations of the NEC.

The NEC is a product of NFPA (The National Fire Protection Association), and therefore the focus of the document is the prevention of fires. Luminaires are heat-producing devices, so this is a genuine concern. Article 410 of the NEC addresses both the installation and the manufacture of luminaires. It contains a lot of useful information with which the designer should become familiar.

NFPA 101 – The Life Safety Code

The Life Safety Code is another NFPA product, so the focus is on personnel safety in the event of a building fire. The code covers several of the most common building types, and specifies the type and amount of exit and emergency lighting required for each. Every lighting design must meet these requirements, so every designer needs to be familiar with them.

ASHRAE/IES Standard 90.1

Standard 90.1 is a US National energy standard developed jointly by the American Society of Heating, Refrigeration, and Air conditioning Engineers (ASHRAE), and the Illuminating Engineering Society of North America (IESNA). It mandates the maximum energy consumption allowable for HVAC and lighting systems in new construction. Meeting the Standard 90.1 requirements is mandatory for all new US Government construction, and most states have adopted it as well. It is only a matter of time before all building departments adopt the Standard, so lighting designers should begin complying with the standard now.

Standard 90.1 addresses lighting by establishing lighting power budgets for various types of buildings and activities. The interior

Table 2.3. Lighting power densities using the building area method

Building type	*Lighting power density (W/ft²)*
Automotive facility	1.5
Convention center	1.4
Court house	1.4
Dining: bar lounge/leisure	1.5
Dining: cafeteria/fast food	1.8
Dining: family	1.9
Dormitory	1.5
Exercise center	1.4
Gymnasium	1.7
Hospital/health care	1.6
Hotel	1.7
Library	1.5
Manufacturing facility	2.2
Motel	2.0
Motion picture theater	1.6
Multi-family	1.0
Museum	1.6
Office	1.3
Parking garage	0.3
Penitentiary	1.2
Performing arts theater	1.5
Police/fire station	1.3
Post office	1.6
Religious building	2.2
Retail	1.9
School/university	1.5
Sports arena	1.5
Town hall	1.4
Transportation	1.2
Warehouse	1.2
Workshop	1.7

(Reprinted by permission from ASHRAE Standard 90.1-1999. Copyright 1999, American Society of Heating, Refrigerating and Air-Conditioning Engineers, Inc. Contact ASHRAE at 404-636-8400 or www.ashrae.org to purchase the standard in its entirety)

lighting power allowance may be determined for the building as a whole, or on a space-to-space basis. Figure 2.17 lists the allowable lighting power density for various types of buildings.

In calculating power usage either within a building or a space, the total allowable luminaire wattage may be found by multiplying the square footage of the space times the allowable power density. When using the space-by-space calculation method, the total building installed lighting power is the sum of the power calculated for all the spaces.

Certain types of lighting are exempt from the allowable power calculations. Specifically, these are:

1. display lighting in museums, galleries, and monuments.

Building Type	Office—Enclosed	Office—Open Plan	Conference Meeting/Mulitpurpose	Classroom/Lecture/Training	Audience/Seating Area	Lobby	Atrium—first three floors	Atrium—each additional floor	Lounge/Recreation	Dining Area	Food Preparation	Restrooms	Corridor/Transition	Stairs—Active	Active Storage	Inactive Storage	Electrical/Mechanical	Building Specific Space Types and LPDs (W/ft^2)		Additional Power Allowance (see 9.3.1.2)
	Space-by-Space Method LPDs: Common Space Types and LPDs (W/ft^2)																			
Office Buildings																				
Office	1.5	1.3	1.5	1.6		1.8	1.3	0.2	1.4	1.4	2.2	1.0	0.7	0.9	1.1	0.3	1.3	Banking Activity Area	2.4	✓
																		Laboratory	1.8	✓
Penitentiary Buildings																				
Penitentiary	1.5	1.3		1.4	1.9	1.8			1.4	1.4	2.2	1.0	0.7	0.9	1.1	0.3	1.3	Confinement Cells	1.1	
Religious Buildings																				
Religious Buildings	1.5	1.3	1.5	1.6	3.2	1.8	1.3	0.2	1.4		2.2	1.0	0.7	0.9	1.1	0.3	1.3	Worship-Pulpit, Choir	5.2	✓
																		Fellowship Hall	2.3	
Retail Buildings																				
Retail	1.5	1.3	1.5			1.8	1.3	0.2	1.4	1.4	2.2	1.0	0.7	0.9	1.1	0.3	1.3	General Sales Area	2.1	✓
																		For accent lighting, see 9.3.1.2.1.(c)		
																		Mall Concourse	1.8	✓
Sports Arena Building																				
Sports Arena	1.5	1.3	1.5		0.5	1.8	1.3	0.2	1.4	1.4	2.2	1.0	0.7	0.9	1.1	0.3	1.3	Ring Sports Arena	3.8	
																		Court Sports Arena	4.3	
																		Indoor Playing Field Area	1.9	
Storage Buildings																				
Warehouse	1.5	1.3	1.5			1.8	1.3	0.2				1.0	0.7	0.9	1.1	0.3	1.3	Fine Material Storage	1.6	
																		Medium/Bulky Material Storage	1.1	
Parking Garage	1.5					1.8						1.0	0.7	0.9	1.1	0.3	1.3	Parking Area - Pedestrian	0.2	
																		Parking Area - Attendant only	0.1	
Theater Buildings																				
Performing Arts	1.5				1.8	1.2	1.3	0.2	1.4			1.0	0.7	0.9	1.1	0.3	1.3			✓
Motion Picture					1.3	0.8			1.4			1.0	0.7	0.9	1.1	0.3	1.3			✓
Transportation Buildings																				
Transportation	1.5	1.3	1.5		1.0	1.8	1.3	0.2	1.4	1.4	2.2	1.0	0.7	0.9	1.1	0.3	1.3	Airport - Concourse	0.7	✓
																		Air/Train/Bus - Baggage Area	1.3	
																		Terminal - Ticket counter	1.8	✓

Figure 2.17. Lighting power densities using the space-by-space method. (Reprinted by permission from ASHRAE Standard 90.1-1999. Copyright 1999, American Society of Heating, Refrigerating and Air-Conditioning Engineers, Inc. Contact ASHRAE at 404-636-8400 or www.ashrae.org to purchase the standard in its entirety)

2. lighting integral to equipment, such as medical and food service equipment;
3. lighting for plant growth;
4. lighting in spaces specifically designed for use by the visually impaired;
5. lighting in retail display windows, provided that the display area is enclosed by ceiling height partitions;
6. lighting in interior spaces that have been specifically designated as a registered interior historic landmark;
7. lighting that is an integral part of advertising, directional, or exit signage;
8. lighting that is for sale or lighting educational demonstration systems;
9. lighting for theatrical purposes, including performance, stage, film and video production;
10. lighting for athletic playing areas with permanent facilities for television broadcasting;
11. casino gaming areas.

Normally, you would only use the whole building method for projects where the whole building *is* the space, such as a warehouse, or retail facility. The space-by-space method allows tradeoffs between the spaces, as long as the total building installed power usage does not exceed the interior lighting power allowance for the building type.

In some cases, Standard 90.1 allows an increase in the interior power allowance when using the space-by-space method for calculating installed interior lighting power. These increases are:

1. 1 watt per square foot when decorative lighting, such as chandeliers and sconces, are installed for highlighting purposes in addition to the general lighting;
2. 0.35 watts per square foot when required to meet task requirements in video display terminal areas;
3. 1.6 watts per square foot for highlighting general merchandise, and 3.9 watts per square foot for highlighting jewelry in retail facilities.

Standard 90.1 also includes a clause which makes it mandatory that every building larger than 5000 square feet contains an automatic control device to shut off building lighting in all spaces. This device can be controlled by either a timer, an occupancy sensor, or a switch. In addition, each space must contain a control device for the lighting within the space. The Standard also specifies additional control requirements for specialized applications.

Control of the space lighting within the space is standard good practice, and is currently in use throughout the industry. A central control point for all building lighting will require some forethought on the part of the designer.

A 100% adoption of Standard 90.1 by building codes agencies will create some changes in current lighting design practices, and the fledgling designer will do well to become familiar with this Standard.

EPACT 92

EPACT 92, or the national Energy Policy Act of 1992, is a US Government sponsored program to reduce the amount of lighting energy consumed by the country. Where Standard 90.1 restricts the energy consumption of the space, EPACT 92 restricts the energy consumption of the lamps themselves. After November 1, 1995, the manufacture or import of many of the then most popular lamp types was forbidden by the Federal government. Failure to comply with the Act could result in fines of up to $100.00 per lamp. At that rate, getting caught with a truckload of illegal lamps would put most people out of business. Needless to say, the lamp manufacturers have taken this Act to heart, and only those lamps with acceptably high efficacy are available now.

This posed a lamp replacement dilemma for those who already had thousands of fluorescent luminaires installed with ballasts designed for the outlawed low efficacy fluorescent lamps. Lamp manufacturers responded by producing lower wattage lamps that would fit into the same luminaire, and operate on the same ballast. But alas, there is no free lunch, even in lighting, and less watts meant less light output. In installations that were already borderline, this resulted in some dimly lit office space.

There is a great movement toward retrofitting existing luminaires with high efficacy lamps and electronic ballasts in order to get acceptable lighting levels out of existing luminaire installations. We'll discuss this in greater detail in Chapter 6 'The Second Time Around'.

IES recommended footcandles

The IES has, since 1947, published a lighting handbook which recommends ambient footcandle levels for most kinds of interior spaces. Later versions of the handbook take into account light quality as well as quantity, and different categories of illumination have been developed. Such factors as age of occupants, contrast of

Table 2.4. Determination of illuminance categories.* (lx = lux, fc = footcandle.)

Orientation and simple visual tasks. Visual performance is largely unimportant. These tasks are found in public spaces where reading and visual inspection are only occasionally performed. Higher levels are recommended for tasks where visual performance is occasionally important.		
A	Public spaces	30 lx (3 fc)
B	Simple orientation for short visits	50 lx (5 fc)
C	Working spaces where simple visual tasks are performed	100 lx (10 fc)
Common visual tasks. Visual performance is important. These tasks are found in commercial, industrial and residential applications. Recommended illuminance levels differ because of the characteristics of the visual task being illuminated. Higher levels are recommended for visual tasks with critical elements of low contrast or small size.		
D	Performance of visual tasks of high contrast and large size	300 lx (30 fc)
E	Performance of visual tasks of high contrast and small size, or visual tasks of low contrast and large size	500 lx (50 fc)
F	Performance of visual tasks of low contrast and small size	1000 lx (100 fc)
Special visual tasks. Visual performance is of critical importance. These tasks are very specialized, including those with very small or very low contrast critical elements. Recommended illuminance levels should be achieved with supplementary task lighting. Higher recommended levels are often achieved by moving the light source closer to the task.		
G	Performance of visual tasks near threshold	3000–10 000 lx (300–1000 fc)

*Expected accuracy in illuminance calculations are given in Chapter 9 of the IESNA Lighting Handbook, Lighting Calculations. To account for both uncertainty in photometric measurements and uncertainty in space reflections, measured illuminances should be within ±10% of the recommended value. It should be noted, however, that the final illuminance may deviate from these recommended values due to other lighting design criteria. (Source: IESNA Lighting Handbook, 9th edn.)

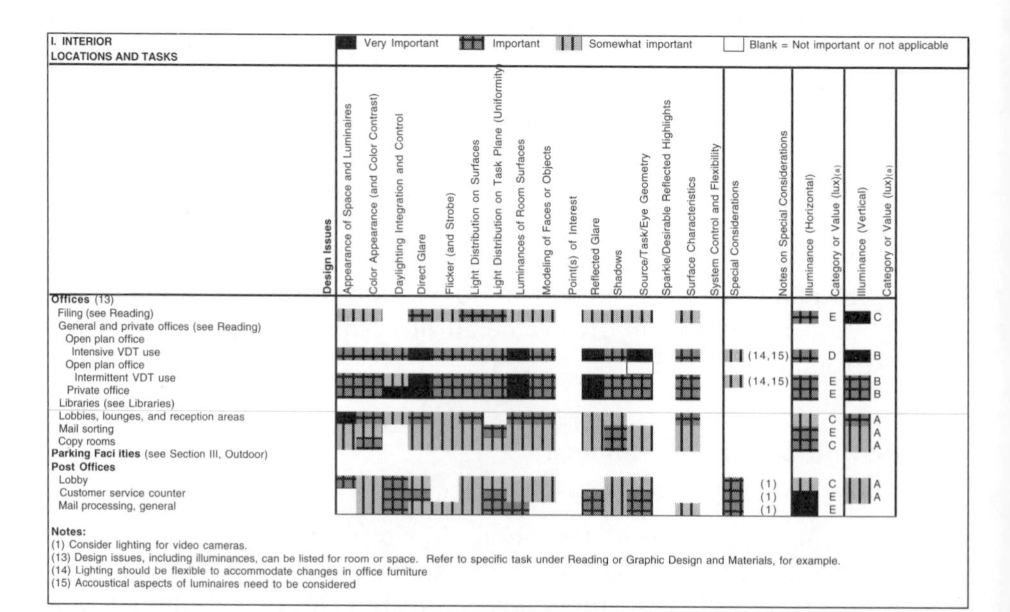

Figure 2.18. IESNA Lighting Design Guide excerpt (source: *IESNA Lighting Handbook*, 9th edn).

tasks, use of video display terminals (VDTs) and degree of detail are weighted into the recommendations for various activities within those spaces. These recommendations are updated periodically to take into account such factors as new technology and energy trends.

These recommendations are the result of a great number of hours of work and research by IES members, and they are recognized industry wide as a design standard. A good lighting designer is well advised to take advantage of the work that has been done, and to use these recommendations as a basis of design. Table 2.4 is an excerpt from the *IES lighting Handbook Reference and Application, Ninth Edition,* which explains the illumination categories of the IES Guidelines. Figure 2.18 shows the IES design guide listings for offices.

Well, OK, now you know enough to select a lamp to provide the proper type and color of light for your design space, and you can select a type of luminaire to put the light into the work area. You can control the luminaires, and you know where to find the proper illumination level for your space and activity. Then you can calculate the number of luminaires required to give you that level. So, armed with this knowledge, you are ready to venture forth into the world of design techniques to try your wings. But first, as always, let's try a few of these fun-filled exercises to refresh our memory.

Exercises

1. What are the two most important methods by which we can produce light by using electricity?
2. How does incandescence produce light?
3. How does photoluminescence produce light?
4. What do spectral energy distribution curves tell us about a light source?
5. What does the CRI number tell us about a light source?
6. Why is an incandescent lamp envelope filled with inert gas?
7. What is the diameter of a PAR30 lamp?
   a. 4 3/4 in.
   b. 3 3/4 in.
   c. 3 in.
   d. 4 in.
8. Approximately what percentage of commercial space in the US is illuminated by fluorescent lamps?
9. A fluorescent lamp with one pin in each end is:
   a. Hot cathode rapid start
   b. Instant start

10. What is the diameter of an F40T12 lamp?
    a. 1.2 in.
    b. 1–1/2 in.
    c. 1 in.
    d. 12 in.
11. What type of lamp is a M250/U
12. What are the two functions of a ballast?
13. You are designing HID lighting for BigBucks Inc.'s new manufacturing facility. A 50 horsepower centrifuge is connected to your panel. What type of ballast should you specify?
    a. R
    b. CW
    c. HX
14. What are two purposes of reflectors?
15. Which type of diffuser would you use in a high abuse area?
    a. Acrylic
    b. Polycarbonate
16. How many lumens will a lamp rated 3000 lumens produce in a luminaire having a ballast factor of 0.85?
17. What is the RCR of a 12 ft × 12 ft room which has a ceiling height of 10 ft, and a working plane of 2 1/2 ft?
18. How many luminaires of the type shown in Figure 2.15 are required to produce 65 fc in the space of exercise 17?
19. You want to turn off the lights in the offices of BigBucks Inc. when nobody is in them. What type of control should you use?
    a. Photoelectric
    b. Occupancy sensors
    c. Dimmers
20. Why do we need to be familiar with building Codes and Standards?

Big bonus exercise

Download the VISUAL Basic Edition Tutorial from the www.lithonia.com/visual Web site. Print it out, and go through the tutorial exercise. This will take you through all the features available in the program. When you have finished the tutorial, print out your results.

3 The design process

A space design is an accumulation of ideas coupled with a knowledge of physical entities directed toward providing a particular environment. The same can be said for lighting design, which plays a major part in accomplishing the goals of the space designer. The design usually begins with the architect's concept of the ambience of the space. This is conveyed to the lighting designer, who makes suggestions for lighting the space based on the architect's ideas, the physical geometry of the space, the intended use of the space, the intended occupants of the space, the budget for the space, and a knowledge of the tools at his disposal.

This is likely to become an iterative process when those grandiose first thoughts bump into the realities of budget constraints. Even so, the design guidelines established at the first meeting with the architect should be followed as closely as possible.

At that first meeting, the architect will present his or her ideas to the designer, along with the floor plans for the space, and (hopefully) elevations and sections showing wall and ceiling construction. If not shown, the designer must ask the architect to define the ceiling height and type for each area within the space. This is usually done on a reflected ceiling plan, which shows the actual ceiling plan and grid layout that the architect intends to use. Wall and floor finishes should also be discussed, since these will affect lighting calculations later on. If not obvious, the intended use for each area should be discussed, as well as furniture layout, the type of tasks to be performed there, and the average age of the expected occupants. Finally, special effects lighting, owner's preferences for types of specialty luminaires, and an order-of-magnitude budget should be discussed. With all of this information firmly in hand, the lighting designer is prepared to take a first cut at designing a lighting system for the space.

Most lighting system designs include four basic types of lighting. These are: *ambient lighting, task lighting, accent lighting*, and *lighting for life safety*. Let's now take a look at each of these four types of lighting.

Ambient lighting

Ambient lighting illuminates the entire space to a level which allows intended tasks to be performed comfortably. This generally means an even distribution of some quantity of light on an imaginary *work plane* selected for the tasks to be performed. The work plane in an office, for example, is 30 in. (76 cm) above the floor, the height of the average desktop.

The ambient light in a space also needs to have the proper quality to support the tasks being performed. In a space with a high use of computers, for instance, minimizing glare is very important. If the tasks being performed are color sensitive, such as matching fabrics or paint, a high color rendering index (CRI) is desired.

Color is also important in creating mood: an efficient office space needs the 'cool', efficient environment created by a 4100 K lamp; a cozy restaurant needs the warm, intimate feel produced by a 2700 K lamp; and public spaces are usually best suited for the neutral color of a 3000 K lamp. Let's now take a specific example and try to develop an approach to ambient lighting design.

1. Luminaire selection

Let's say that we have just been commissioned by an architect to provide the lighting design for a new office/conference room addition to Bullmoose Industries' office building. We sit down with the architect to go over the project requirements. He presents us with the floor plan shown in Figure 3.1.

He explains to us that the office area will have no permanent partitions in it. The workspaces will be separated with movable partitions, and the arrangement will be changed from time to time, as employees move in and out. The architect says that he will select the partitions, so we request that he select a light color, with a reflectance value as high as possible. The architect further informs us that the ceiling will be a suspended grid ceiling, using white acoustical tile, 9 ft above the floor, with a 2 ft by 2 ft grid spacing throughout the addition. With a little prodding, we discover that the average age of the office workers will be about 45 years, and that their work will be with normal contrast black and white text, with a high usage of computer video data terminals (VDT). The architect goes on to tell us that we have a moderate budget for this project, and the client wants to put most of his money into the conference room, which will be seen by prospective clients. The conference room will be multipurpose, and will be used for corporate training and board meetings, as well as for high tech presentations to those

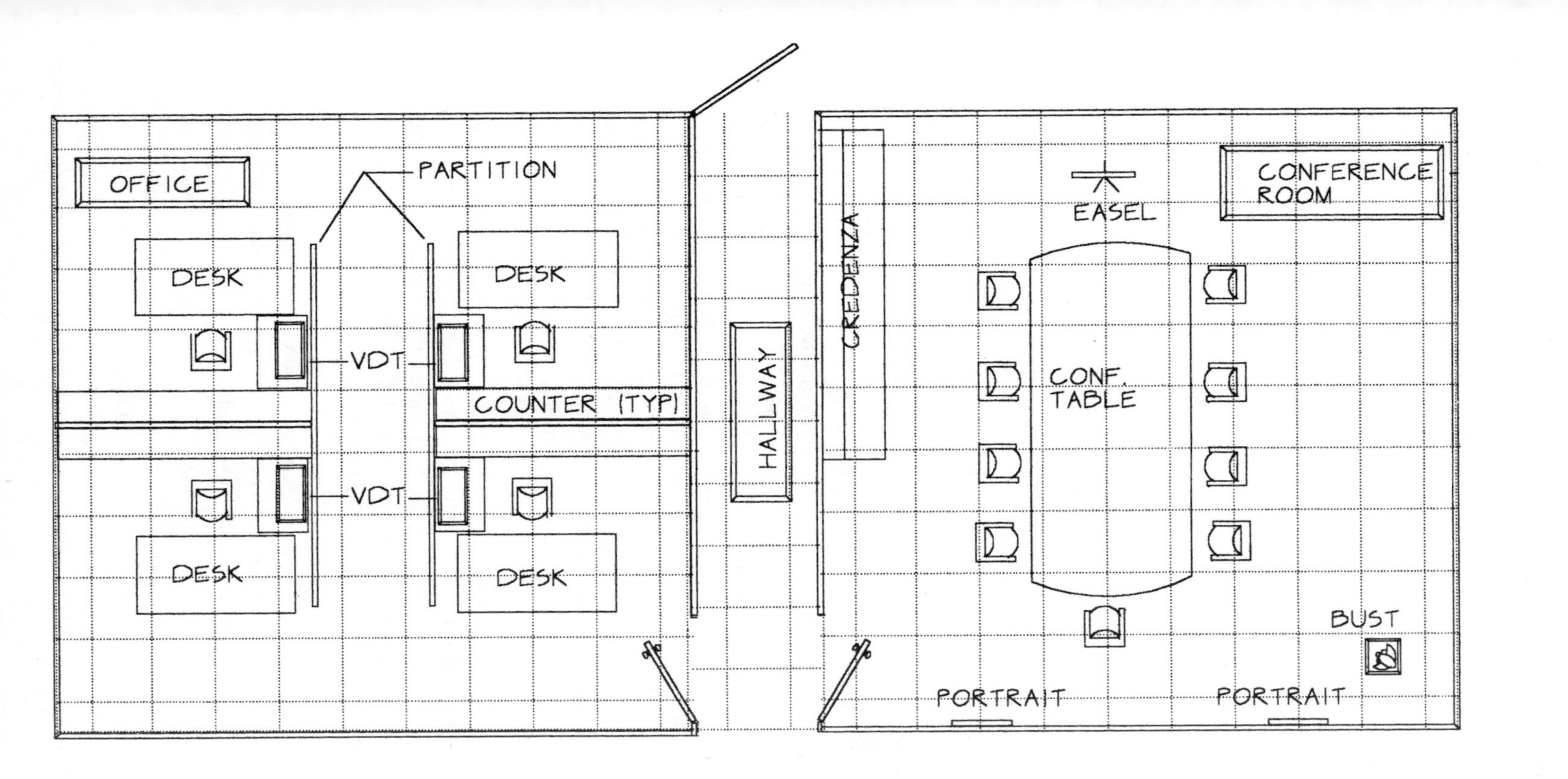

Figure 3.1. Bullmoose office addition floor plan.

prospective clients. The conference room will have two company portraits on the south wall, and a bust on a pedestal in the south-east corner of the room.

With this information in hand, we can start to work. First, we will select our light source for the ambient lighting of the office space. We will use a fluorescent source, since it provides more light for less energy input, and we can choose the color temperature and CRI rating to suit our needs. For this application, a color temperature of 4100 K and a CRI of about 75 will be adequate, since color rendering isn't of primary importance, and the higher CRI lamps are more expensive.

Now we need to select the luminaire which we will be using. An ideal luminaire for the low glare, movable partition environment of the office space would be a suspended indirect luminaire, since indirect lighting would be uniform throughout the space and there would be no glare. Unfortunately, our budget and low ceiling will not allow the use of suspended luminaires, so we are left with troffers which fit into the ceiling grid. Now we need to find out what is available, and what the costs are.

Here's where we will call in the person who is going to be one of our closest allies during the design: the manufacturer's representative. In general, manufacturers' representatives are highly knowledgeable sales professionals who have a greater in-depth understanding of their products than most designers can get by skimming through catalogs. Many representatives also have well-developed design skills. A lot of time can be saved, and a lot of good ideas can be brought to the table by involving the manufacturer's representatives early in the design.

How you find these reps is relatively easy: if you know the brand name of a luminaire that you want to use, you can go to www.lightsearch.com and search by manufacturer to get their Web address. Then, simply go to their Web site, do a little clicking, and locate the representative nearest you. If you have no idea of what brand will suit your needs, you can search the lightsearch site by product description (i.e. 'fluorescent troffers'), and find out who makes what you need, and then go to their Web site. For a general listing of lighting manufacturers, you can go to www.lighting-inc.com/searchman.html. If all else fails, you can always to go to a Web search engine such as www.yahoo.com, and type in the generic description of what you want (i.e. 'fluorescent lighting') and start visiting the Web sites returned from your search until you find a manufacturer who builds what you want. Then locate the representative as above.

When you contact your local rep, it helps to remember that representatives are salesmen first and foremost, so they usually represent a full line of luminaires and associated equipment. They probably have an offering for every luminaire type that you need, and they are happy to provide you with manufacturers' literature and product application guidance. In return, they expect the opportunity to bid their equipment on your project.

Let's say that our local rep has been found, has stopped by with a few softback quick selector catalogs, and we are ready to select luminaires. We sit with the rep and look together at the criteria for ambient lighting of the office space:

1. the system should have low glare, because of the high use of VDTs;
2. the system should provide even light distribution throughout the space, because of the owner's desire for flexibility in furniture placement;
3. the system should provide 'cool' light, in order to get an efficient 'feel' within the space; color rendering is of secondary importance;
4. the system should provide adequate light for workers over the age of 45 to perform tasks comfortably;
5. as always, the system should comply with the energy requirements of ASHRAE 90.1;
6. the system should fit within a moderate budget – not 'cheap', but not 'gold plated', either.

When the lighting rep reviews these criteria, he will likely suggest a fluorescent troffer with either a parabolic diffuser or a direct/indirect diffuser which approximates the suspended indirect luminaire. Fluorescent is a good choice, for reasons mentioned previously. Troffers are suitable because of ceiling type, and the flexibility in arrangement to provide good light distribution. Either parabolic or direct/indirect diffusers will reduce glare. Now, if we take a little time to reflect on the salient features of the parabolic diffuser mentioned in Chapter 2, we will realize that parabolics are not ideal for our application because of the sharp light cutoff that they provide. Cutoff is beneficial in an open plan office, but when partitions are introduced into the space, light below the cutoff angle is blocked by the partitions, as shown in Figure 3.2. This prevents an even contribution of light from all luminaires to the working plane.

We will go with the direct/indirect luminaire to get the best distribution possible.

We now need to select a direct/indirect luminaire that will fit into the 2 ft by 2 ft ceiling grid of the office space. Let's say that we happen to have the Lithonia lighting catalog. Lithonia's softback

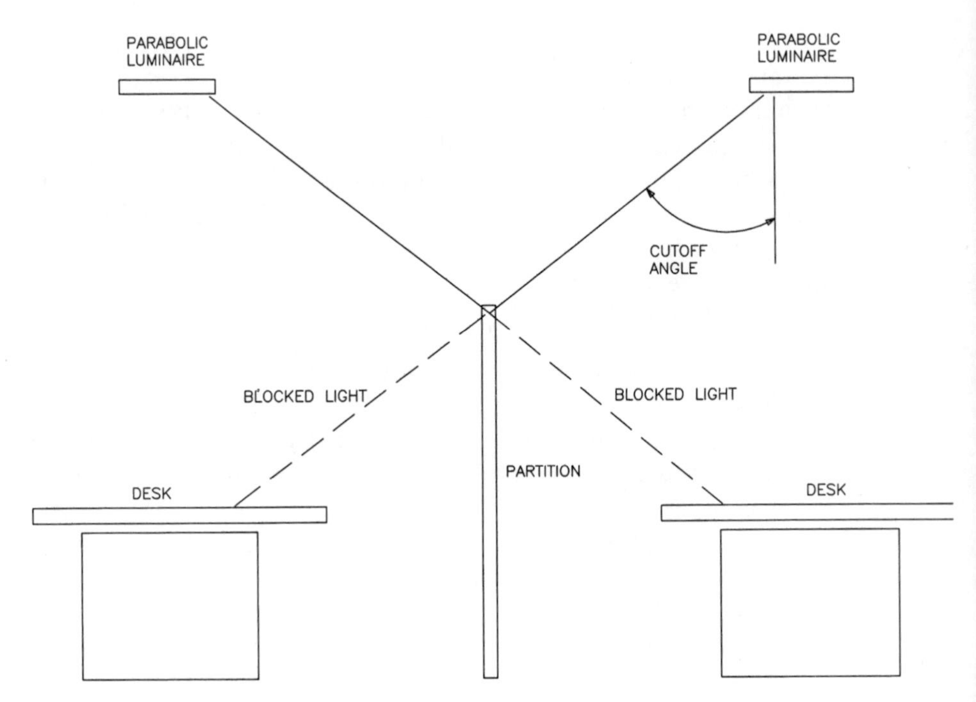

Figure 3.2. Parabolic luminaire distribution blocked by partition.

catalog is called the Product Selection Guide, and in it we find their recessed direct/indirect luminaire, which is called the 'Avante' luminaire, type AV. We see from the catalog that we have a choice of either 2 ft × 4 ft, or 2 ft × 2 ft type AV luminaire that will fit easily into the grid. Figure 3.3 is the catalog sheet for the type AV. To determine which size will best meet our performance criteria, we need to see the photometric distribution curves of each luminaire for comparison.

The Product Selection Guide is like most manufacturer's softback catalogs, and doesn't contain all the photometric distribution curves. We can either request that the rep send us the full data sheets for the type AV, or, we can go online and download them ourselves. Since we're in the twenty-first century, and have computers and all, let's download the data.

If you remember, in Chapter 2, we downloaded Lithonia's VISUAL BASIC program and photometric data to perform some calculations. Let's go back to Lithonia's central Web address, www.lithonia.com, and see what we can find. When we get to the opening screen, we see that there is a Product Info selection, which will take us to the Online Catalog. We can click through the Indoor and Fluorescent sections, until we find Recessed AV, which is what we're looking for. This will work for us, but it will take quite a bit

FEATURES

OPTICAL SYSTEM

- Twin matte white polyester powder paint finished reflectors provide uniform light distribution. Optional diffuse aluminum stepped reflectors available.
- All diffusers control direct light distribution and glare by shieding lamps from direct view.
- All shieldings snap into place by pivoting on light trap for easy lamp access.
- Injection molded light traps prevent light leaks between shielding and endplates.

SHIELDING OPTIONS

- Metal Diffuser staggered Round holes (MDR) 52% open perforated metal with .075" diameter holes backed with white acrylic diffuser.
- Straight Blade Louver (SBL) sides of perforated metal with staggered round holes and solid blade louvered center. Sides and louver backed with white acrylic diffuser.
- Metal Diffuser aligned Mini slots (MDM) 46% open perforated metal backed with white acrylic diffuser.
- Acrylic Diffuser Prismatic lens (ADP) extruded acrylic lens backed with white acrylic diffuser.
- Metal Diffuser staggered Linear slots (MDL) 45% open perforated metal backed with white acrylic diffuser.

ELECTRICAL SYSTEM

- Class P, Thermally protected, resetting, HPF, Non-PCB, UL Listed, CSA-certified electromagnetic ballast is standard. Energy saving and electronic ballast are sound rated A. Standard combinations are CBM approved and conform to UL 935.

HOUSING

- Housing is powder painted cold rolled steel. All edges hemmed or rounded.
- Trims available for standard 1" tee bar, mini-tee bar, screw slot grids.
- Drywall ceiling adapters available.
- Fixtures can be row mounted end to end.

LISTING

- UL listed and labeled. Listed and labeled to comply with Canadian and Mexican Standards (see options).

Specifications subject to change without notice.

Catalog Number	Type

Direct/Indirect General Lighting System

AV 2'x4'

T8, T5 or T5HO
1 or 2 lamp
Compact Fluorescent
1 lamp in cross section

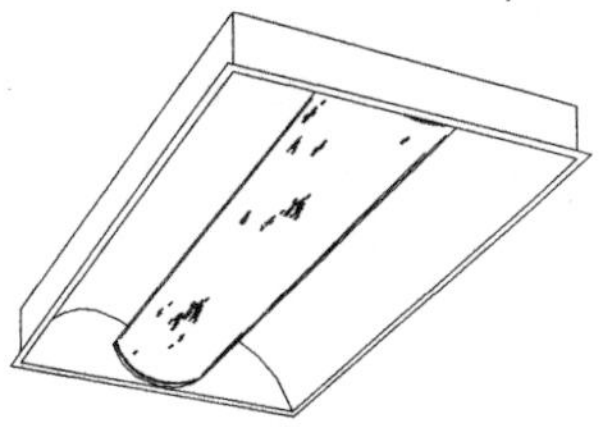

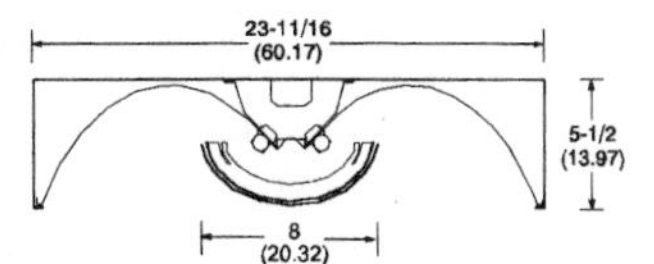

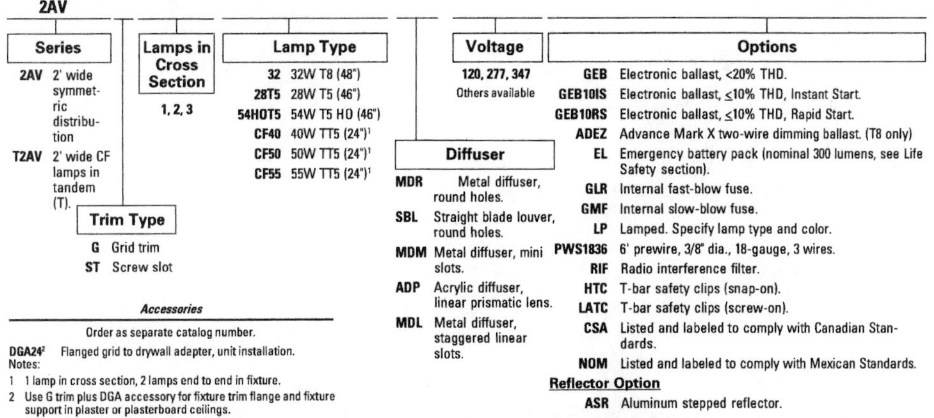

ORDERING INFORMATION

Example: **2AV G 2 32 MDR 120 GEB**

2AV

Series

2AV	2' wide symmetric distribution
T2AV	2' wide CF lamps in tandem (T).

Lamps in Cross Section

1, 2, 3

Trim Type

G	Grid trim
ST	Screw slot

Lamp Type

32	32W T8 (48")
28T5	28W T5 (46")
54HOT5	54W T5 HO (46")
CF40	40W TT5 (24")[1]
CF50	50W TT5 (24")[1]
CF55	55W TT5 (24")[1]

Diffuser

MDR	Metal diffuser, round holes.
SBL	Straight blade louver, round holes.
MDM	Metal diffuser, mini slots.
ADP	Acrylic diffuser, linear prismatic lens.
MDL	Metal diffuser, staggered linear slots.

Voltage

120, 277, 347
Others available

Options

GEB	Electronic ballast, <20% THD.
GEB10IS	Electronic ballast, ≤10% THD, Instant Start.
GEB10RS	Electronic ballast, ≤10% THD, Rapid Start.
ADEZ	Advance Mark X two-wire dimming ballast. (T8 only)
EL	Emergency battery pack (nominal 300 lumens, see Life Safety section).
GLR	Internal fast-blow fuse.
GMF	Internal slow-blow fuse.
LP	Lamped. Specify lamp type and color.
PWS1836	6' prewire, 3/8" dia., 18-gauge, 3 wires.
RIF	Radio interference filter.
HTC	T-bar safety clips (snap-on).
LATC	T-bar safety clips (screw-on).
CSA	Listed and labeled to comply with Canadian Standards.
NOM	Listed and labeled to comply with Mexican Standards.

Reflector Option

ASR	Aluminum stepped reflector.

Accessories

Order as separate catalog number.

DGA24[2]	Flanged grid to drywall adapter, unit installation.

Notes:

1 1 lamp in cross section, 2 lamps end to end in fixture.

2 Use G trim plus DGA accessory for fixture trim flange and fixture support in plaster or plasterboard ceilings.

LITHONIA LIGHTING

AV 2x4

Figure 3.3. Lithonia type AV luminaire (source: Lithonia Lighting).

of time to evaluate all the photometric options for the AV. Since we're looking for comparative photometrics, there's an easier way to go. Back at the opening screen, we see a Lighting Software section. Go there. You'll see, among the options, a Photometric Viewer Download. We want to download this viewer. It's a few megabytes, so the download may take a little while. Once the download is finished, you'll need to unzip the file, and install the program according to the instructions. Now you're ready to go.

When you bring up the viewer, you'll see a folder called Lithonia Photometrics, Open it, then go to the Fluorescent folder. Open it, and go to the Architectural folder. Here you'll see the AV folder. Open it, and it brings up all the photometric files available for the AV luminaire. We're looking for a comparison between a 2 ft × 2 ft and a 2 ft × 4 ft luminaire, so we choose a luminaire with two CF40 lamps for the 2 ft × 2 ft, and one with three 32 W lamps for the 2 ft × 4 ft luminaire. Use the Set Compare button, select the two lamps, and hit the View button. When the photometric summary comes up, select the CP Curve option. You'll see the comparative photometrics for the two luminaires. This should look like Figure 3.4.

We see two curves, one looking down the axis of the lamps, and one looking perpendicular to the lamp axis. You can see from Figure 3.4 that the distribution pattern for both luminaires is symmetrical, regardless of viewing angle. This means that if either of the luminaires were installed, and all the desks were rotated 90°, the lighting levels at the desktop would remain the same.

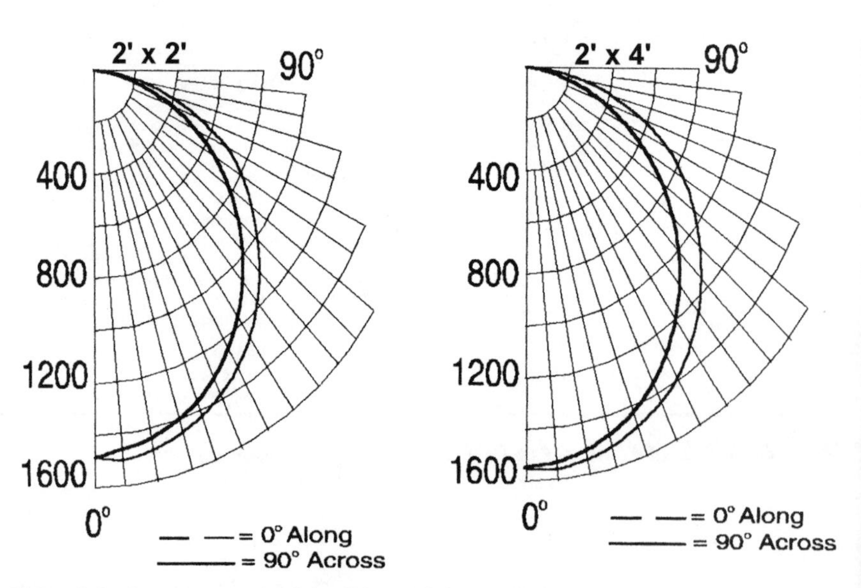

Figure 3.4. Comparison of 2 ft × 2 ft and 2 ft × 4 ft luminaire photometric curves (source: author/Lithonia Lighting).

Either is a good choice, then, since we need flexibility of furniture placement and orientation. Now what lamp should we use?

To make our final selection, we have to investigate further. Certainly energy consumption will be one of our main concerns, since we will have to meet ASHRAE 90.1 energy constraints with our design. We also don't want to encumber the client with high energy bills, or waste the country's valuable energy resources. In this vein, let's compare the lamps available to us, and see where our best value is.

Looking at the catalog information, we see that the 2 ft × 2 ft luminaire will accept the 40 W compact T5, the 50 W compact T5 and the 55 W compact T5. The 2 ft × 4 ft luminaire will accept the F32T8 lamp, the F28T5 lamp, and the F54T5 HO lamp. We have a choice of the number of lamps.

If our lighting rep has brought us a lamp catalog, we can look up each lamp to see which has the best efficacy, or lumens per watt in our color range. If we don't have a lamp catalog, we can go back to the Internet, and pull up a lamp manufacturer's Web site. Several lamp manufacturers have their product catalogs online, and if you have a reasonably recent browser, you're all set. Let's go to the Philips site, www.philips.com.

When we get to the opening screen, we see that we have a multitude of information at our fingertips, some useful, and some not applicable to our project. We click on the Professional Product heading at the top of the screen, and select Lighting, because that's what we're interested in. Under lighting, we see catalogs, and we have a choice of their European or North American offerings. We'll select North American, and we see that there is a Full Lighting Product Catalog available. This is Philips' on-line catalog. Click on that. We can download the entire catalog, and have it on hand for future projects, or we can just retrieve the information we need, and get on with the job. For now, we'll just get what we need. Under the Advanced Search section, in the 'Lamps' window, select compact fluorescent, and click on search. Philips' entire line of compact fluorescent lamps will be returned. Scroll through offerings down to 50W, and you'll see several 50W PL-L long fluorescent lamps. That's what we want. The 'long' indicates that the lamp is 24 inches long. Each offering is for a different color temperature. If you select the first lamp, you'll see that it is 4100K, which is what we want, with an initial lumen output of 4300 lumens, and a CRI of 82. The listing also includes 'design lumens', which is the initial lumens multiplied by the Lamp Lumen Depreciation (LLD). The 50W lamp appears to fit our needs. Now, in order to find the 40W lamp, we'll refine our search efforts somewhat.

Back-up to the search menu and again select compact fluorescent. This time, pick the 'watts' scroll arrow, and scroll down to 40W. Philips doesn't offer a 40W lamp. The closest to it is a 38W lamp. Select that, and only the 38W offerings are returned. Select the first one, and you'll see that it has an output of 3300 lumens, a color temperature of 4100K, and a CRI of 82. Also a good choice. Now go back to the search menu, select Fluorescent, 32W, and T8 for tube type. This zeros right in on the lamp we are looking for. We select the 4100K offering, and find that it has an initial lumen output of 2950 lumens, and a CRI of 86. Another winner. We could go through the same process and obtain the data for our T5 lamps, the 28WT5 and the high output, F54T5 HO, but the T5 lamps are expensive, and not readily obtainable in some places, so we'll deem them not suitable for our purposes. T5 lamps are very efficient, however, and should be considered for projects which will be maintained professionally. One thing to remember is that the high output F54T5 HO lamps produce so many lumens (5000), that they can only be used in indirect luminaires, where the source is not visible.

Now we have all the lamp data we need to see which lamp best fits our needs, so let's look at the efficacies (lumens per input watt ratio) for each of our selected lamps using initial lumens.

For the 50W compact fluorescent, efficacy is 4300 lumens/50W=86; and for the compact 40WT5, we have 3300 lumens/40W=82.5, and for the F32T8, efficacy is 2950 lumens/32W=92. Clearly, the lamp with the best efficacy in this lot is the F32T8 lamp, so we would like to use F32T8's if the design will allow. Any of these lamps will operate efficiently on an electronic ballast, which we will use to eliminate flicker.

Now we have two luminaires complete with lamps and ballast that we can use to provide ambient lighting for the office space. The question is, how many of each of these luminaires will we need, and where should they be placed?

To determine the number of luminaires, we first need to determine the number of footcandles we need on the desktops in the space. This requires going to the IES-recommended footcandle tables and looking up our space. The proper classification for our space is open plan office with intermittent VDT use. If we look back at Figure 2.18, we see that our office falls into category E, which corresponds to a value of 50 footcandles (50 fc). For a first cut at determining quantities of each type of luminaire, we can run some zonal cavity calculations. If you don't remember the zonal cavity calculation procedure from Chapter 2, now would be a good time to take a couple of minutes for review. Or. . . we can run our zonal cavity calculations using that nifty photometric viewer program that we just downloaded. For those of you who enjoy grinding through the math by hand, have fun. You have all the data needed to evaluate a 2 ft × 4 ft luminaire with two, and three lamps, and a 2 ft × 2 ft luminaire with two 40 W lamps. Our office space is 20 ft × 20 ft, with a 9 ft ceiling, and a work plane 2.5 ft from the floor.

O.K. – just to humor the math whizzes among you, here's how the zonal cavity calculation would go for the 2 ft × 2 ft luminaire with two CF40W lamps:

$$\text{Room cavity ratio, or RCR} = \frac{5(6.5) \times (20\text{ ft} + 20\text{ ft})}{(20\text{ ft} \times 20\text{ ft})} = 3.5$$

We can pull up the photometric viewer, and find our coefficient of utilization (CU) is 0.5. We use a generic electronic ballast factor of 0.80, and since an office is a relatively clean environment, a luminaire dirt depreciation (LDD) of 0.90. That gives us a total light loss factor (LLF) of 0.9 × 0.8, or 0.72. Use the design lumens of 2970 to account for LLD, and for the desired 50 fc, we have:

$$\text{Luminaires} = \frac{50 \times (20\text{ ft} \times 20\text{ ft})}{2(\text{lamps}) \times 2970 \times 0.72 \times 0.5} = 9.35$$

For the rest of us, let's look back at the photometric viewer and click on the AV luminaire. Then let's select the 2AVG2CF40ADP luminaire. That's the one with two CF40 lamps. Next, hit the View button to bring up the photometric summary. This time, we will select the Room Estimator option. The estimator asks us to fill in the room dimensions, and the luminaire mounting height. The viewer provides default working plane height, lamp lumens, and light loss factor. The RCR, and CU are also provided by the program, having been calculated from room data input and stored luminaire data. We can type in the desired footcandles at 50, and the program runs a zonal cavity calculation to show that we need... What? 6.2 luminaires? Why is this so different from the 9.35 luminaires that we calculated by hand? Let's go back and look at the estimator screen. The default LLF is 1.0, which doesn't take into account the ballast factor, or the LDD, which brings our LLF down to 0.72. We type that in, and see that we still only need 8.7 luminaires. The program also defaults to 3100 lumens/lamp, which is lamp initial lumens. If we go back and plug in the design lumens of 2970, we see that the required luminaires goes to 9.2, which is close enough.

This is a good point to remember. When using quick calculator computer programs, always check to make sure that the default values make sense. We know that the LLF is *always* less than 1.0, and that we use design lumens to assure better system performance over the long run.

Now let's run the program for the 2 ft × 4 ft luminaire, with both two and three F32T8 lamps. The program will retain our inputs, so all we need to change is the lamp photometrics, and the LLF to a more reasonable value of 0.65. After a bit of mousework, we see that we need 9.4 of the two-lamp luminaires, and 6.7 of the three lamp.

If cost was our major concern, the quick estimator would suffice, and at this point we would choose to go with six of the three lamp luminaires. But we have light distribution as a major concern, so we will now have to run point-by-point calculations to see how our three choices perform. The easiest way to do this, of course, is by using the computer, so now we will pull up the VISUAL BASIC program that we downloaded in Chapter 2.

We will enter the geometry of the room, the ceiling grid type, use standard reflectances, and select the photometric file for the luminaires and lamps that we have chosen. Below the photometric file selector is a light loss factor selector which offers either standard or IES calculated light loss factors. None of the standard values fit our needs, so click on the icon beside the 'LLF Value' window. For the 40 W compact fluorescent lamp, we will use 'fluorescent' and 'T5 Long CF'. For the

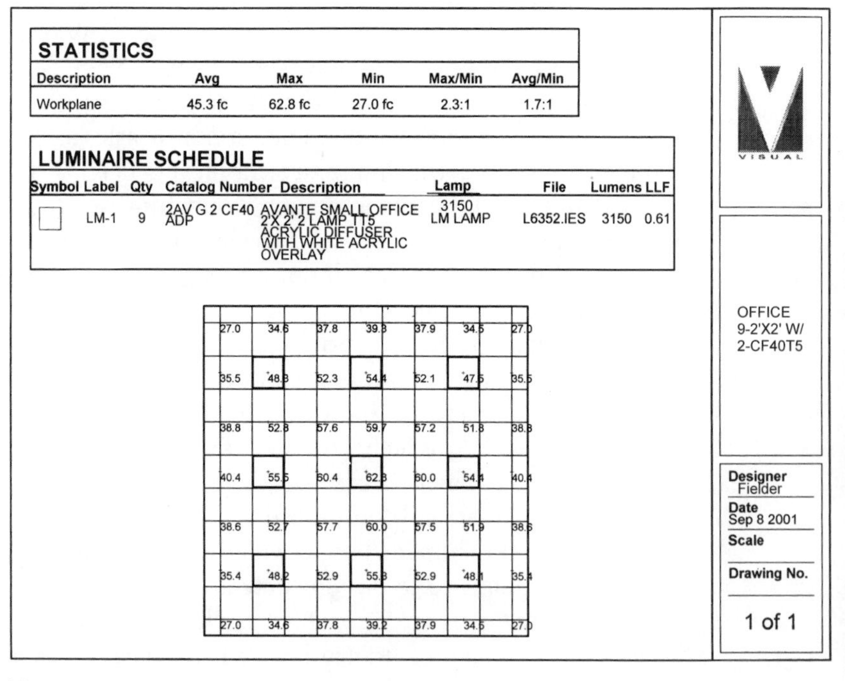

(a)

Figure 3.5. Bullmoose office VISUAL results using (a) 9.2CF40 luminaires, (b) 9.2F32T8 luminaires, and (c) 6.3F32T8 luminaires.

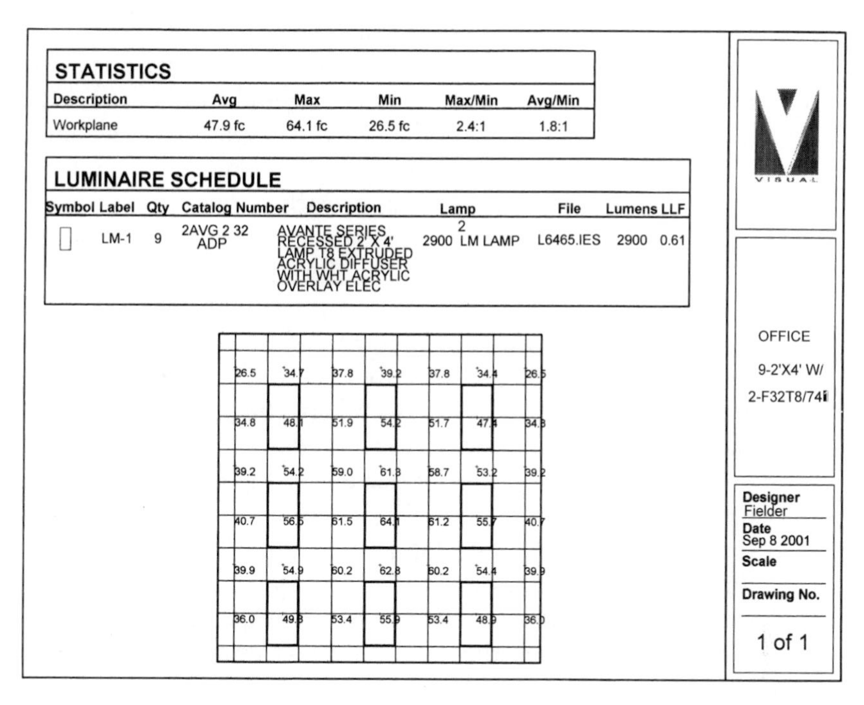
STATISTICS
Description Avg Max Min Max/Min Avg/Min
Workplane 47.9 fc 64.1 fc 26.5 fc 2.4:1 1.8:1
LUMINAIRE SCHEDULE
Symbol Label Qty Catalog Number Description Lamp File Lumens LLF
LM-1 9 2AVG 2 32 ADP AVANTE SERIES RECESSED 2' X 4' LAMP T8 EXTRUDED ACRYLIC DIFFUSER WITH WHT ACRYLIC OVERLAY ELEC 2 2900 LM LAMP L6465.IES 2900 0.61
26.5 34.7 37.8 39.2 37.8 34.4 26.5
34.8 48.1 51.9 54.2 51.7 47.4 34.8
39.2 54.2 59.0 61.3 58.7 53.2 39.2
40.7 56.5 61.5 64.1 61.2 55.7 40.7
39.9 54.9 60.2 62.8 60.2 54.4 39.9
36.0 49.8 53.4 55.9 53.4 48.9 36.0
VISUAL
OFFICE
9-2'X4' W/
2-F32T8/74
Designer
Fielder
Date
Sep 8 2001
Scale
Drawing No.
1 of 1

(b))

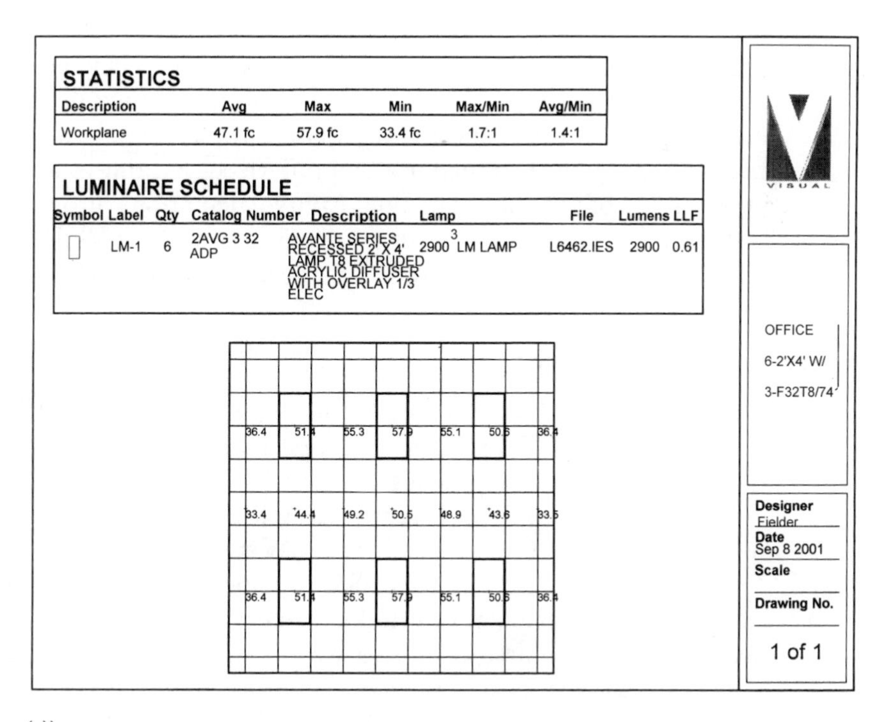
STATISTICS
Description Avg Max Min Max/Min Avg/Min
Workplane 47.1 fc 57.9 fc 33.4 fc 1.7:1 1.4:1
LUMINAIRE SCHEDULE
Symbol Label Qty Catalog Number Description Lamp File Lumens LLF
LM-1 6 2AVG 3 32 ADP AVANTE SERIES RECESSED 2' X 4' LAMP T8 EXTRUDED ACRYLIC DIFFUSER WITH OVERLAY 1/3 ELEC 3 2900 LM LAMP L6462.IES 2900 0.61
36.4 51.4 55.3 57.9 55.1 50.6 36.4
33.4 44.4 49.2 50.5 48.9 43.6 33.5
36.4 51.4 55.3 57.9 55.1 50.6 36.4
VISUAL
OFFICE
6-2'X4' W/
3-F32T8/74
Designer
Fielder
Date
Sep 8 2001
Scale
Drawing No.
1 of 1

(c))

F32T8s we will use 'Fluorescent' and 'T8 Rapid Start'. Both lamps have a calculated LLF of 0.9. Continue on through the windows: For LDD we will use 'Indirect Luminaires'; and for the ballast we will use 'Electronic'. None of the 'other' factors apply to our situation. We are now left with a combined LFF of 0.61. Now run the calculator using a desired illumination of 50 fc for each luminaire.

As the results come up for each luminaire type, make a note of the calculated power density for later comparison with the ASHRAE 90.1 allowances. Make a calculation run for each of the three types, using the 'Tools' menu, and the 'Lumen Method' button to return to the calculation program to change lamps. The program will retain the other settings. When you have finished, the results should look like Figure 3.5a–c.

The first thing that we notice about the results is that the average footcandles computed for the six 3F32T8 luminaires is more than that for the nine 2CF42 luminaires, and about equal to that of the nine 2F32T8 luminaires. The max/min ratio is seen to be better for the six three-lamp luminaires, which indicates a better light distribution. The higher the ratio, the less uniform the light, and vice versa. Further examination shows that the six three lamp luminaires provide approximately 50 fc in the central three-quarters of the room, where the work will most likely take place. This is a good distribution of light, and the lamps with the highest efficacy are being utilized. We will use the six 2 ft × 4 ft luminaires with three F32T8 lamps each. It is interesting to note again, that at no point on the grid do we get 47.1 fc, which was our average illumination, and what we would expect from a zonal cavity calculation. As you become more accustomed to computerized lighting calculations, you will probably abandon zonal cavity calculations altogether, and instead utilize the more advanced programs, such as VISUAL PRO, which will allow you to model spaces much more accurately.

Energy code compliance

OK, now that we have settled in on luminaires for the office, let's run a quick ASHRAE 90.1 check to make sure that the power density is within the guidelines.

From Figure 2.17 in Chapter 2, we can see that our power density allowance for this space is 1.3 W per square foot, or 520 W for the whole 400 sq. ft space. We can look back at the notes we took during the calculations and see that the power density for the six 3F32T8 luminaires is 1.3, so we are just within the guidelines. Our notes show that the other two options exceed 1.3 W/sq ft, and are not within the guidelines. We have made a good choice.

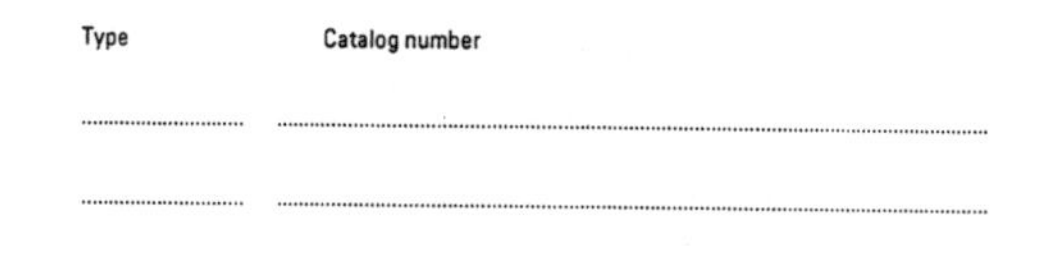

Specification Premium Static Troffer

SP 2'x2'

Compact Fluorescent

2, 3 or 4 lamps

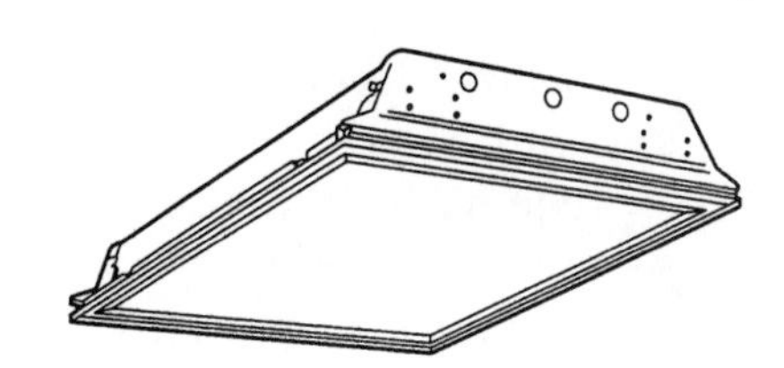

FEATURES

- Full family of specification grade static luminaires offers choice of sizes, ceiling trims, door frames and other options to provide general illumination for every recessed application.
- Standard door is fully gasketed flush steel with mitered appearance — completely frames shielding. Corners screwed together for rigidity, easy lens replacement.
- Urethane foam gasket eliminates light leaks between door frame and housing.
- Overlapping flange and modular ceiling trims factory installed with standard swing-gate hangers or field convertible with optional trim and hanger kits.
- Aluminum door frames available — flush or regressed.
- Integral T-bar safety clips hold T-bar securely, no fasteners required to install.
- Optional spring-loaded latches
- Guaranteed for one year against mechanical defects in manufacture.

SPECIFICATIONS

BALLAST — Thermally protected, resetting, Class P, HPF, non-PCB, UL listed, CSA certified ballast is standard. Energy-saving and electronic ballasts are sound rated A. Standard combinations are CBM approved and conform to UL 935.

WIRING & ELECTRICAL — Fixture conforms to UL 1570 and is suitable for damp locations. AWM, TFN or THHN wire used throughout, rated for required temperatures.

MATERIALS — Housing formed from cold-rolled steel. Acrylic shielding material 100% UV stabilized. No asbestos is used in this product.

FINISH — Five-stage iron-phosphate pretreatment ensures superior paint adhesion and rust resistance. Painted parts finished with high-gloss, baked white enamel.

LISTING — UL listed and labeled. CSA certified (see options). NOM labeled (see options).

Specifications subject to change without notice.

PHOTOMETRICS

Calculated using the zonal cavity method in accordance with IESNA LM41 procedure. Floor reflectances are 20%. Lamp configurations shown are typical. Full photometric data on these and other configurations available upon request.

2SP 2 CF40 A12
Report LTL 3606
S/MH (along) 1.2 (across) 1.3

Coefficient of Utilization

Ceiling	80%			70%			50%		
Wall	70%	50%	30%	70%	50%	30%	50%	30%	10%
1	75	72	70	73	71	68	68	66	64
2	69	64	60	67	63	59	61	58	55
3	64	58	53	62	57	52	55	51	48
4	59	52	47	58	51	46	49	45	42
5	54	46	41	53	46	41	44	40	36
10	37	28	23	36	28	23	27	22	19

Zonal Lumens Summary

Zone	Lumens	%Lamp	%Fixture
0-30	1395	22.5	33.2
0-40	2255	36.4	53.7
0-60	3587	57.9	85.4
0-90	4202	67.8	100.0
90-180	0	0	0
0-180	4202	67.8	100.0

2SP 3 CF40 A12
Report LTL 3334
S/MH (along) 1.2 (across) 1.3

Coefficient of Utilization

Ceiling	80%			70%			50%		
Wall	70%	50%	30%	70%	50%	30%	50%	30%	10%
1	69	66	64	67	65	63	62	60	59
2	63	59	55	62	58	54	56	53	50
3	59	53	48	57	52	48	50	47	44
4	54	47	43	53	47	42	45	41	38
5	50	42	37	49	42	37	41	36	33
10	34	26	21	33	25	21	25	20	17

Zonal Lumens Summary

Zone	Lumens	%Lamp	%Fixture
0-30	1901	20.4	32.9
0-40	3076	33.1	53.3
0-60	4926	53.0	85.3
0-90	5772	62.1	100.0
90-180	0	0	0
0-180	5772	62.1	100.0

2SP 3 U31 A12125
Report LTL 3365
S/MH (along) 1.2 (across) 1.4

Coefficient of Utilization

Ceiling	80%			70%			50%		
Wall	70%	50%	30%	70%	50%	30%	50%	30%	10%
1	69	67	65	68	65	63	63	61	60
2	64	60	56	63	59	55	57	54	52
3	59	54	50	58	53	49	51	48	45
4	55	49	44	54	48	43	46	42	39
5	51	44	38	49	43	38	42	37	34
10	34	26	21	33	26	21	25	21	18

Zonal Lumens Summary

Zone	Lumens	%Lamp	%Fixture
0-30	1745	20.8	33.2
0-40	2871	34.2	54.6
0-60	4631	55.1	88.1
0-90	5257	62.6	100.0
90-180	0	0	0
0-180	5257	62.6	100.0

SP 2x2 CF

Figure 3.6. Lithonia type SP luminaire (source: Lithonia Lighting).

Now let's look at the hallway, which also requires ambient lighting. Hallways, or paths of egress, are required by code to have illumination of 10 fc at the floor. No visual tasks are being performed there, so glare and color rendition are of minimal concern. Ten footcandles doesn't represent many watts/square foot, so if we have reasonable efficacy, we're not likely to get into trouble with the energy code. This means that a 'plain Jane' luminaire will suffice. Looking back at our catalog, we see the 2 ft × 2 ft troffer shown in Figure 3.6.

This luminaire has an acrylic diffuser, and will fit nicely into our ceiling grid. We want to use a 3000 K lamp here, to produce a 'neutral' feel to the space and we'll use the 40 W compact fluorescent lamp because of the efficacy. To calculate the number of luminaires required to produce the 10 fc, we can return to the computer. This time, the working plane is at 0 in. (the floor). Running the program, we find that we can get 10 fc with only one luminaire. This would make our hallway look like a cave, so we'll try two luminaires and see what happens. The VISUAL results are shown in Figure 3.7. The light distribution is fairly even, and we have more than the required 10 fc, so we'll consider this a good fit, and move on.

Now let's take a look at task lighting for the conference room.

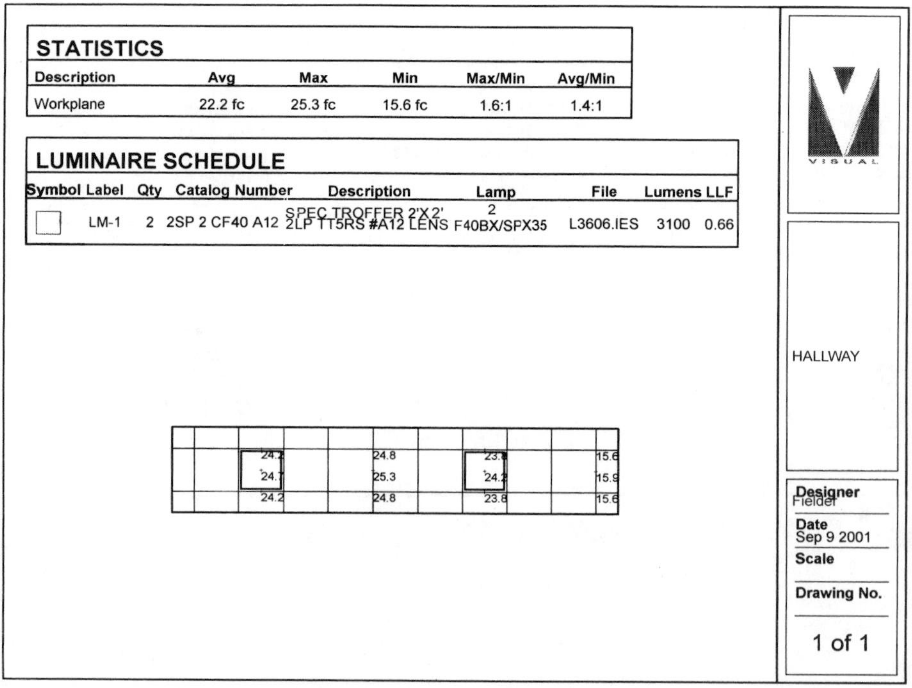

STATISTICS

Description	Avg	Max	Min	Max/Min	Avg/Min
Workplane	22.2 fc	25.3 fc	15.6 fc	1.6:1	1.4:1

LUMINAIRE SCHEDULE

Symbol	Label	Qty	Catalog Number	Description	Lamp	File	Lumens	LLF
☐	LM-1	2	2SP 2 CF40 A12	SPEC TROFFER 2'X2' 2LP TT5RS #A12 LENS	2 F40BX/SPX35	L3606.IES	3100	0.66

Figure 3.7. Bullmoose hallway VISUAL results.

Task lighting

Task lighting means putting the light where you need it to do the job. This infers that task lighting is almost always direct lighting. If we look at the conference room in Figure 3.1, we can see a couple of areas where task activity will take place: the conference table, of course, where everybody will be sitting; and the side table area, where people will refill their coffee cups from time to time. Now, how should we light these areas?

If we look back at our criteria for the conference room, we see that we need enough light on the table to be able to perform detailed work in training sessions (60–70 fc), but we need only enough light to take notes (10–20 fc) when a presentation video is in use. The same is true for the side table: when everybody is up getting coffee and pastries, they need to see what they're doing; when the video is on, you only need minimum light in that part of the room.

This situation will be accommodated by putting dimming controls on the luminaires that we use for the table and the side table. Now what luminaires should we use?

Luminaire selection

Again, for color temperature, CRI, and amount of light produced, we should consider fluorescent first for the lighting of the conference table. We will mount the luminaires directly over the conference table, since that is where the tasks will be performed. The spacing of the luminaires should offer even light distribution on the entire surface of the table, and glare should be minimized. There are no partitions involved in this application, so parabolic diffusers should work well – perhaps better than the direct/indirect luminaires, because light trespass into areas other than the task area will be limited by the cutoff characteristics of the parabolic diffusers.

This time, instead of calculating a number of luminaires required to provide a specific light level, we will locate our luminaires directly above the task area, and space them evenly to provide good light distribution. If we look at the geometry of the conference table, we see that three 2 ft × 4 ft luminaires will fit nicely into the ceiling grid above the table. The long axis of the luminaires will be parallel to the line of vision of those sitting at the sides of the table, which will further reduce glare for the majority of those at the table. If necessary, we can adjust the maximum light output of the luminaires by choosing the number of lamps in the luminaires. We now go back to the catalog to select the luminaire. Figure 3.8 shows the catalog data for an appropriate parabolic luminaire selection.

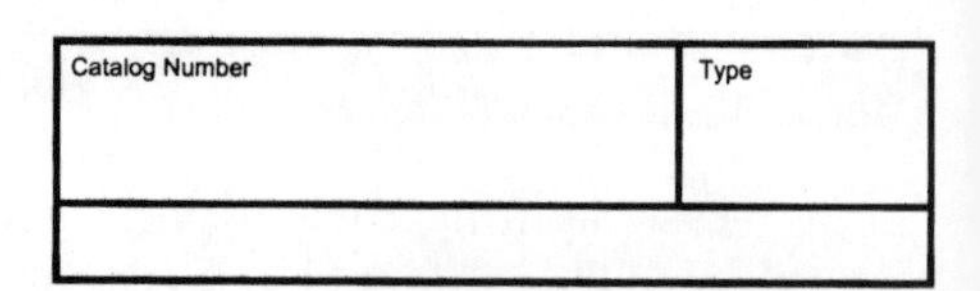

FEATURES

- Choice of low iridescent diffuse or specular louver finishes. Ideal for use with triphosphor lamps.
- Black reveal provides floating louver appearance, conceals optional air-supply slots.
- Overlapping flange and modular ceiling trims factory installed with standard swing-gate hangers or field convertible with optional trim and hanger kits.
- Optional heat-removal dampers and air-pattern control blades allow airflow control.
- T-hinges die-formed for maximum strength. Latches spring-loaded, concealed in reveal.
- Guaranteed for one year against mechanical defects in manufacture.

SPECIFICATIONS

BALLAST — Thermally protected, resetting, Class P, HPF, non-PCB, UL listed, CSA certified ballast is standard. Energy-saving and electronic ballasts are sound rated A. Standard combinations are CBM approved and conform to UL 935.

WIRING & ELECTRICAL — Fixture conforms to UL 1570 and is suitable for damp locations. AWM, TFN or THHN wire used throughout, rated for required temperatures.

MATERIALS — Housing formed from cold-rolled steel. Louvers formed from anodized aluminum. No asbestos is used in this product.

FINISH — Five-stage iron-phosphate pretreatment ensures superior paint adhesion and rust resistance. Painted parts finished with high-gloss, baked white enamel.

LISTING — UL listed and labeled. Listed and labeled to comply with Canadian and Mexican Standards (see Options).

Specifications subject to change without notice.

PARAMAX® Parabolic Troffer

PM4 2'x4'

4" Deep Louver

3 lamps

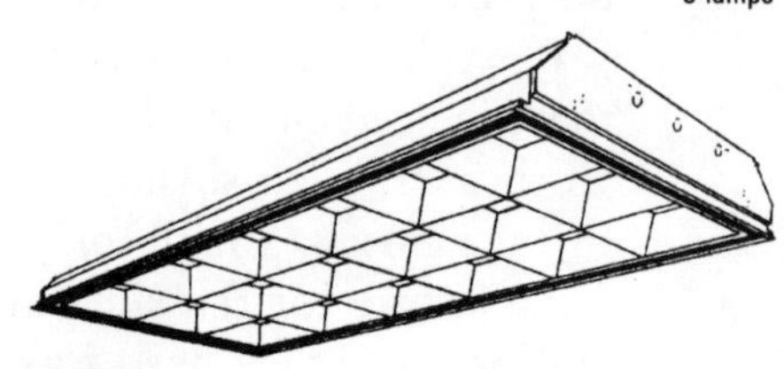

ENERGY

Luminaire Efficacy Rating (LER) and Annual Energy Cost/1,000 Lumens

3-lamp, LD louver - LER.FP = 57. Annual energy cost = $4.21.

Based on 32W T8 lamp, 2850 lumens and electronic ballast.

Ballast factor = .88 and input watts = 86.

Calculated in accordance with NEMA standard LE-5.

PHOTOMETRICS

Calculated using the zonal cavity method in accordance with IESNA LM41 procedures. Floor reflectances are 20%. Lamp configurations shown are typical. Full photometric data on these and other configurations available upon request.

2PM4 G B 3 32 18LD

Report LTL 6757 - Lumens per lamp = 2850

S/MH (along) 1.2 (across) 1.5

Coefficient of Utilization

Ceiling	80%			70%			50%			0%
Wall	70%	50%	30%	70%	50%	30%	50%	30%	10%	0%
0	78	78	78	76	76	76	73	73	73	66
1	73	71	68	71	69	67	66	65	63	58
2	68	63	60	66	62	59	60	57	55	51
3	62	57	52	61	56	51	54	50	47	45
4	58	51	46	56	50	45	48	44	41	39
5	53	46	41	52	45	40	44	40	36	34
6	49	42	36	48	41	36	40	35	32	30
7	46	38	33	45	37	32	36	32	29	27
8	43	35	29	42	34	29	33	29	26	24
9	40	32	27	39	31	27	31	26	23	22
10	37	29	24	37	29	24	28	24	21	20

Zonal Lumens Summary

Zone	Lumens	%Lamp	%Fixture
0-30	1911	22.3	34.1
0-40	3247	38.0	57.9
0-60	5313	62.1	94.8
0-90	5604	65.5	100.0
90-180	0	0	0
0-180	5604	65.5	100.0

PM4 2x4 3LP

Figure 3.8. Lithonia Type PM4 luminaire (source: Lithonia Lighting).

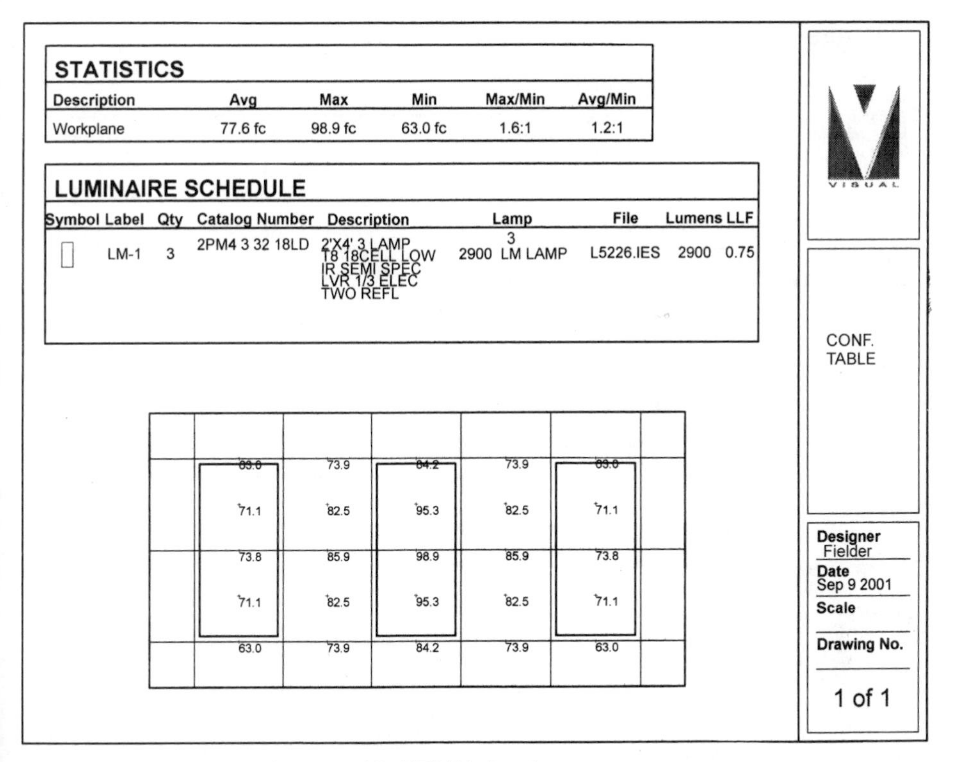

STATISTICS

Description	Avg	Max	Min	Max/Min	Avg/Min
Workplane	77.6 fc	98.9 fc	63.0 fc	1.6:1	1.2:1

LUMINAIRE SCHEDULE

Symbol	Label	Qty	Catalog Number	Description	Lamp	File	Lumens	LLF
▯	LM-1	3	2PM4 3 32 18LD	2'X4' 3 LAMP T8 18CELL LOW IR SEMI SPEC LVR 1/3 ELEC TWO REFL	3 2900 LM LAMP	L5226.IES	2900	0.75

Figure 3.9. Bullmoose conference table VISUAL Results.

A zonal cavity calculation for the area above the table wouldn't have much meaning, but we can run the computer analysis for that area. We assume that our space stops at the boundaries of the table, and that the wall reflectance is zero. This time we use the feature of the program which allows us to place the luminaires where we want them. First, let's try three F32T8/841 lamps per luminaire. Figure 3.9 shows the results of the computer run.

We can see that three lamps per luminaire will give us about 99 fc in the center of the table, and an average of 78 fc. This is about right for maximum illumination for close work at the table. We're going to use dimming ballasts on these luminaires, so we can reduce the level for less intensive tasks. Now, what about the side table?

Let's take a look at how we want the side table lighting to function: it must provide enough light to operate a coffee pot; it must be dimmable to reduce brightness during presentations; and we also don't want to be able to see the source of light while seated at the conference table. Because of the room layout, we can see that troffers aren't suitable for this job. Looking back into our toolbox from Chapter 2, we can see that a recessed downlight will give us direct lighting, and also hide the light source. We must go back to our

FEATURES

OPTICAL SYSTEM

- Reflector - Self-flanged, specular clear, highly specular or semi-diffuse reflector.
- Baffle/cone - Specular clear upper reflector. Microgroove baffle with white painted flange or specular black cone with flange that matches cone finish.
- Brightness control and high efficiency are optimally balanced.
- Controlled anodized coating suppresses iridescence.

HOUSING

- 16-gauge galvanized steel mounting/plaster frame with friction support springs to retain optical system. Maximum 7/8" ceiling thickness.
- Expandable, self-locking mounting bars provide horizontal and vertical adjustment.
- Galvanized steel junction box with bottom-hinged access covers and spring latches. Two combination 1/2"–3/4" and four 1/2" knockouts for straight-through conduit runs. Capacity: 8 (4 in, 4 out) No. 12 AWG conductors, rated for 75°C.

ELECTRICAL SYSTEM

- Die-cast aluminum socket housing. Ventilated top for convective cooling.
- Horizontally-mounted, positive-latch, thermoplastic socket(s) in single (centered) or double-lamp configuration.
- Class P (thermally protected) high power factor ballast(s) mounted to the junction box.

LISTING

- Fixtures are UL listed for thru-branch wiring, recessed mounting and damp locations. Listed and labeled to comply with Canadian Standards (see Options).

Compact Fluorescent Downlights

6" AF

Open Reflector

Horizontal Lamp
Twin-Tube, Double Twin-Tube or Tri-Tube

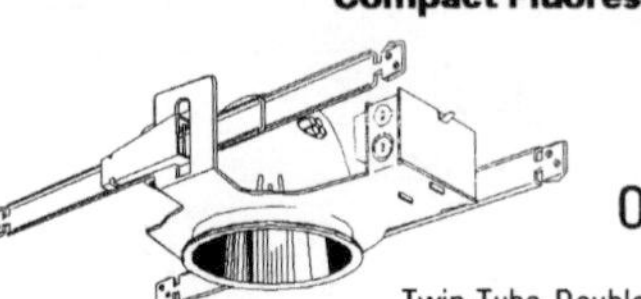

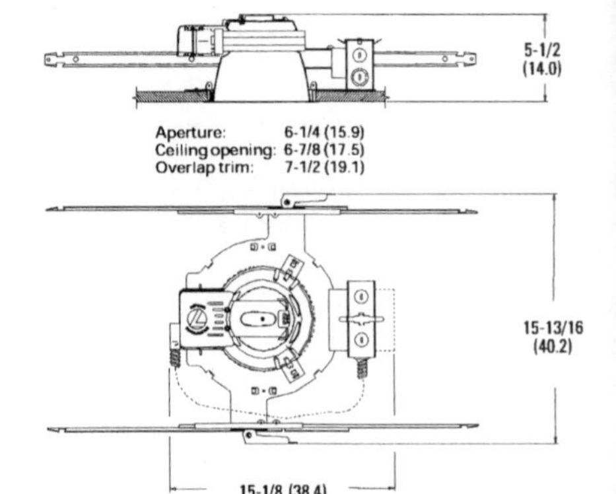

All dimensions are inches (centimeters).
Maximum dimension for two ballast configuration. Actual configuration may vary.

ENERGY

LER.DOL	Annual Energy Cost	Lamps	Lamp Lumens	Ballast Factor	Input Watts
29	$8.23	2/26DTT	3600	.98	52

Calculated in accordance with NEMA standard LE-5.

ORDERING INFORMATION

Example: **AF 2/18DTT 6AR 120 GEB10 WLP**

Choose the boldface catalog nomenclature that best suits your needs and write it on the appropriate line. Order accessories as separate catalog numbers (shipped separately).

AF

Series

- **AF**

Wattage/Lamp

- **1/9TT**[1] One 9W twin-tube
- **2/9TT**[1] Two 9W twin-tube
- **1/13TT**[1] One 13W twin-tube
- **2/13TT**[1] Two 13W twin-tube
- **1/13DTT**[2] One 13W double twin-tube
- **2/13DTT**[2] Two 13W double twin-tube
- **1/18DTT**[2] One 18W double twin-tube
- **2/18DTT**[2] Two 18W double twin-tube
- **1/26DTT**[2] One 26W double twin-tube
- **2/26DTT**[2] Two 26W double twin-tube
- **1/18TRT**[3] One 18W tri-tube
- **1/26TRT**[3] One 26W tri-tube
- **1/32TRT**[3] One 32W tri-tube
- **1/42TRT**[3] One 42W tri-tube

Reflector Type

- **6AR** Clear
- **6PR** Pewter
- **6UBR** Umber
- **6WTR** Wheat
- **6CR**[4] Champagne Gold
- **6GR**[4] Gold
- **6BC**[5] Black Cone
- **6MB**[5] Black Micro-groove Baffle

Finish

- **(blank)** Specular low iridescent
- **LD** Semi-diffuse low iridescent
- **LS** Highly specular

Voltage

- **120**
- **277**
- **347**

Ballast[6]

- **EMB** Electromagnetic ballast. Requires two-pin lamp.
- **GEB10** Electronic ballast. Requires four-pin lamp.
- **DMHL** Lutron Hi-lume® electronic dimming ballast. (120V or 277V; 18DTT, 26DTT, 32TRT and 42TRT only.) Requires four-pin lamp. Minimum dimming level 5%.
- **ADEZ** Advance Mark X electronic dimming ballast. (120V or 277V; 26DTT, 26TRT, 32TRT and 42TRT only.) Requires four-pin lamp. Minimum dimming level 5%.

Options

- **WLP** With 35K lamp (shipped separately).
- **TRW** White painted flange. (Standard with 6MB.)
- **LRC**[7] Provides compatibility with Lithonia Reloc System. Lithonia Reloc System can be installed **less this option** with connectors provided by others. Access above ceiling required.
- **GMF** Single slow-blow fuse
- **RIF** Radio interference filter.
- **ELR**[8] Emergency battery pack. Remote test switch provided.
- **EL**[8,9] Emergency battery pack. Integral test switch provided.
- **QDS**[10] Quick disconnect for easy ballast replacement.
- **DS** Dual switching.
- **GSKT** 1/8" x 3/8" foam gasketing.
- **CSA** Listed and labeled to comply with Canadian Standards.

Accessories

Order as separate catalog numbers.

- **SC6** Sloped ceiling adaptor. Degree of slope must be specified (10D, 15D, 20D, 25D, 30D). Ex: SC6 **10D**.
- **CTA6** Ceiling thickness adaptor. Not available with TRT fixtures. (Extends mounting frame to accommodate ceiling thickness up to 2".)

NOTES

1. Available with electromagnetic ballast only.
2. Available with electromagnetic or electronic ballast.
3. Available with electronic ballast only.
4. Not recommended for use with compact fluorescent lamps; consult factory.
5. Not available with finishes.
6. Refer to options and accessories tab for additional ballast types.
7. For compatible Reloc systems, refer to options and accessories tab.
8. For dimensional changes, refer to options and accessories tab.
9. Available in two-lamp units only.
10. Not available with ELR or EL option.

100-AF6

Figure 3.10. Lithonia type AF luminaire (source: Lithonia Lighting).

catalog and find a downlight which suits our purpose. The catalog specification sheet for one such luminaire is shown in Figure 3.10.

A word of caution here: when selecting recessed downlights, always find out whether insulation is to be installed above the ceiling in which the luminaire is mounted. If it is, you must use an 'IC' (insulation contact) rated luminaire, or you will have excessive heat buildup.

Fortunately, the ceiling in our conference room has no insulation above it, so we can use a non-IC rated luminaire. We will use a fluorescent lamp, so that we may match the color of the lamps over the conference table, and we will use a dimming ballast to dim the luminaire when desired. Horizontal lamp mounting will be selected, so that the lamp will not be visible when the luminaire is viewed from an angle. To further camouflage the light source, we will use a black grooved baffle trim on the inside of the luminaire.

Our selected luminaire offers a 42 W triple tube compact fluorescent lamp as an option, so we will use this to get as much undimmed light as possible out of the luminaire. No particular footcandle level is required here, so three luminaires, mounted directly above the side table, should suffice.

Even though there is no furniture shown on the east wall, there is a big open space there, so there's a pretty good chance a telephone table, or something else, will eventually be put there. We will put two downlights over that area just for good measure.

We now have the task areas lit, so what else do we need? If we look around, we see a presentation easel in the front of the room, the company portraits on the back wall, and the bust of Commodore Bullmoose himself in the back corner. It would be nice to accent these features with light, so let's see what we can do.

Accent lighting

Accent lighting, like task lighting, is direct lighting projected on a specific area for a specific purpose: *Task lighting illuminates an area so that a task may be performed there; accent lighting illuminates an object so that it may be seen clearly and dramatically.*

The majority of accent lighting that we encounter in the real world is done from above, with ceiling-mounted luminaires. Most of these luminaires, particularly in retail spaces, are track mounted. Track lights are versatile, easily adjusted, and easily re-lamped to perform a variety of accent lighting jobs. In our conference room, however, the architect feels that a surface-mounted track would 'clutter' his ceiling, and take away from the effect that he is trying

FEATURES

OPTICAL SYSTEM

- Internal housing components painted matte black. Lamp snoot minimizes stray light in housing.
- Self-flanged, specular clear, semi-diffuse or highly specular anodized cone maximizes output while minimizing room-side flash.
- Seamless cast faceplate is retained by self-aligning, constant force, torsion support springs.
- Accommodates up to two lens and/or louver filters.
- 0°-45° vertical adjustment and 360° horizontal adjustment.
- Tool-less vertical and horizontal lamp adjustments are made with optical system lowered below ceiling for simple focusing. Adjustment mechanism is lockable to maintain focus during relamping.
- Locking mechanism is visible from below ceiling with trim assembly lowered.
- Position indicators allow consistent aiming from fixture to fixture.
- Softening lens standard.

MECHANICAL SYSTEM

- Re-lamp capability from above or below ceiling.
- Tool-less removal of step-down transformer.
- Tool-less removal of thermally-activated insulation detector.
- Universal housing with matte black finish and plaster flange will accommodate a wide variety of DLV series trims. Maximum 2" ceiling thickness.
- Expandable, self-locking adjustable mounting bars standard.
- Secondary housing adjustment system for precise, final ceiling to flange alignment.
- Painted steel junction box with bottom-hinged access covers and spring latches.

ELECTRICAL SYSTEM

- Replaceable socket assembly. MR16 socket assembly standard (20W - 75W).
- Four combination 1/2"-3/4" knockouts for straight-through conduit runs. Capacity: 8(4 in, 4 out) No. 12 AWG conductors, rated for 90°C.

LISTING

- UL listed for thru-branch wiring, recessed mounting and damp locations.
- Listed and labeled to comply with Canadian Standards.

Low Voltage Downlights

3" DLV

Adjustable Downlight

Seamless Cast Faceplate with Open Cone

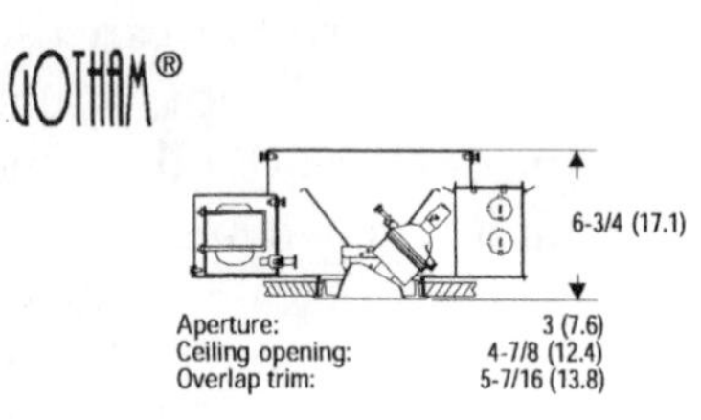

All dimensions are inches (centimeters).

13-3/4 (34.9)

16-1/8 (41.0)

ORDERING INFORMATION

Example: **DLV ADJ MR16 3BCT30 120 DWHG**

Choose the boldface catalog nomenclature that best suits your needs and write it on the appropriate line. Order accessories as separate catalog numbers (shipped separately).

DLV ADJ

Series

DLV	

Configuration

ADJ	

Lamp designation

MR16	MR16 capability (Standard)
MR11	MR11 capability
ALR12	ALR12 capability
ALR18	ALR18 capability
AR70[1]	AR70 capability

Cone/Color

3BC[2]	Black
3AC	Clear
3PC	Pewter
3WTC	Wheat
3UBC	Umber
3GC	Gold
3CC	Champagne gold
3MB[2]	Black baffle

Type

T30	Tapered cut, for angles 25° - 45°
T20	Tapered cut, for angles 15° - 25°
T00	Staight cut, for angles 0°-15°

Finish

(blank)	Specular low irides-cent
LD	Semi-diffuse low irides-cent
LS	Highly specular

Volt

120
277
347

Options

LRC[3]	Provides compatibility with Lithonia Reloc System. Lithonia Reloc System can be installed **less this option** with connectors provided by others. Access above ceiling required.
SHHSG[4]	Accommodates 3" adjustable trim. HSG height reduced to 5-3/4" (14.6). HSG not compatible with other trim types.

Architectural Colors (powder finish)[5]

Standard Colors

DWHG	Matte white (standard)
DDB	Dark bronze
DBL	Black
DWH	Gloss white

Classic Colors

DMB	Medium bronze
DNA	Natural aluminum
DSS	Sandstone
DGC	Charcoal gray
DTG	Tennis green
DBR	Bright red
DSB	Steel blue

Accessories

- For lens accessories, refer to options and accessories tab.
- For lamp options, refer to options and accessories tab.

NOTES

1. Clear safety lens standard with AR70 lamp designation.
2. Not available with finishes.
3. For compatible Reloc systems, refer to options and accessories tab.
4. Not available with ALR18 and AR70 lamps.
5. Colors available for faceplate. Additional architectural colors available; please see brochure 794.3.

LITHONIA LIGHTING

Figure 3.11. Lithonia Type DLV luminaire (source: Lithonia Lighting).

to achieve. We must go back to the catalogs to find a recessed, adjustable downlight that will accommodate the type of lamps that we require for accent lighting. The catalog data for one such luminaire is shown in Figure 3.11.

The luminaire shown in Figure 3.11 is available in adjustable accent and wallwash accent configurations. It is designed to utilize the MR-16 low voltage lamp, which, as you should remember from Chapter 2, produces a brilliant, focused white light in several beam spreads. It is also very hot, but as said before, we have no insulation concerns in our space.

The rule of thumb for wallwashing in general is to space the luminaires equally from the wall and each other to obtain uniform illumination. In our case, we are using the wallwashers for accent lighting, so we will mount a wallwash luminaire in front of each of the portraits, so that the light strikes the portrait at an angle of 30° or less from the vertical. This is to prevent glare from the glass covering of the portraits. The mechanics of this is illustrated in Figure 3.12.

This will give us an oval of light on the portrait for an accent effect. The lamp position and beam spread are both selectable, and may be adjusted to frame the entire portrait. For our presentation easel, we will mount one of the accent downlights so that we can frame the entire presentation surface and maintain our 30° angle. For our pièce de résistance, we will locate two downlights over the bust of the Commodore, to accent his chiseled features and give

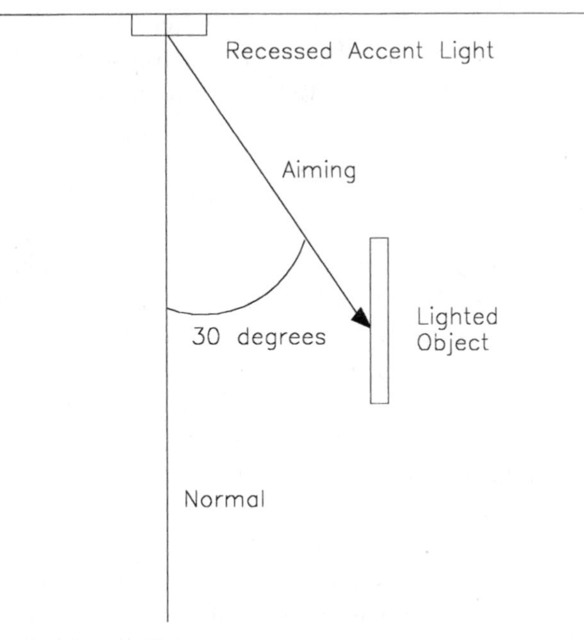

Figure 3.12. Accent luminaire mounting.

him a three-dimensional appearance (remember 'light and shadows' from Chapter 1). Again, these downlights should be placed so that the light hits the sculpture at an angle of 30° or less from the vertical to eliminate glare from the Commodore's prominent proboscis. All of the accent luminaires will be adjusted after installation to maximize their effect.

Finally, we will put dimmers on the accent lights so that they may be dimmed while presentations are taking place, and that about wraps it up for the new spaces – or does it?

What happens if the electricity goes off just when Bullmoose, Inc. has a conference room full of foreign investors? Those guys will be crashing around in the dark, trying to find their way out of unfamiliar surroundings, and not feeling very benevolent toward Bullmoose, Inc., unless. . . unless we have done a good job of lighting for life safety.

Lighting for life safety

Lighting for life safety means providing enough light for occupants to safely exit a building in case of a power outage, which often accompanies a fire in the building. It also entails guiding those occupants out of the building by the use of exit signs.

The Life Safety Code (NFPA 101) requires that we provide an average of 1 fc at the floor on the paths of egress from the building in emergency conditions. Power for the emergency lighting can either be provided by a backup battery system, or an emergency generator with an automatic switch to transfer loads from the normal to the generator source. Battery systems can be purchased in two configurations: a centralized battery, charger, and transfer switch that powers a number of emergency luminaires; or individual batteries, charger, and transfer switch located within the luminaire itself. Most fluorescent luminaire manufacturers offer the emergency backup option, up to 1100 lumens per luminaire, as a standard package.

The Bullmoose building has no emergency generator, so that option is out for us. A centralized battery system is best suited for a large area in which a number of emergency luminaires are required, but would not be cost-effective for our small area, which will use only a few emergency luminaires. We are left with providing battery backup in a few selected luminaires in our space. Now, which luminaires should we choose?

Emergency lighting design is about 90% 'feel', and 10% calculation, if any calculation at all is required. Since 1 fc is very easy to obtain, we can place luminaires equipped with emergency packs almost anywhere within our small spaces and get there. In our case,

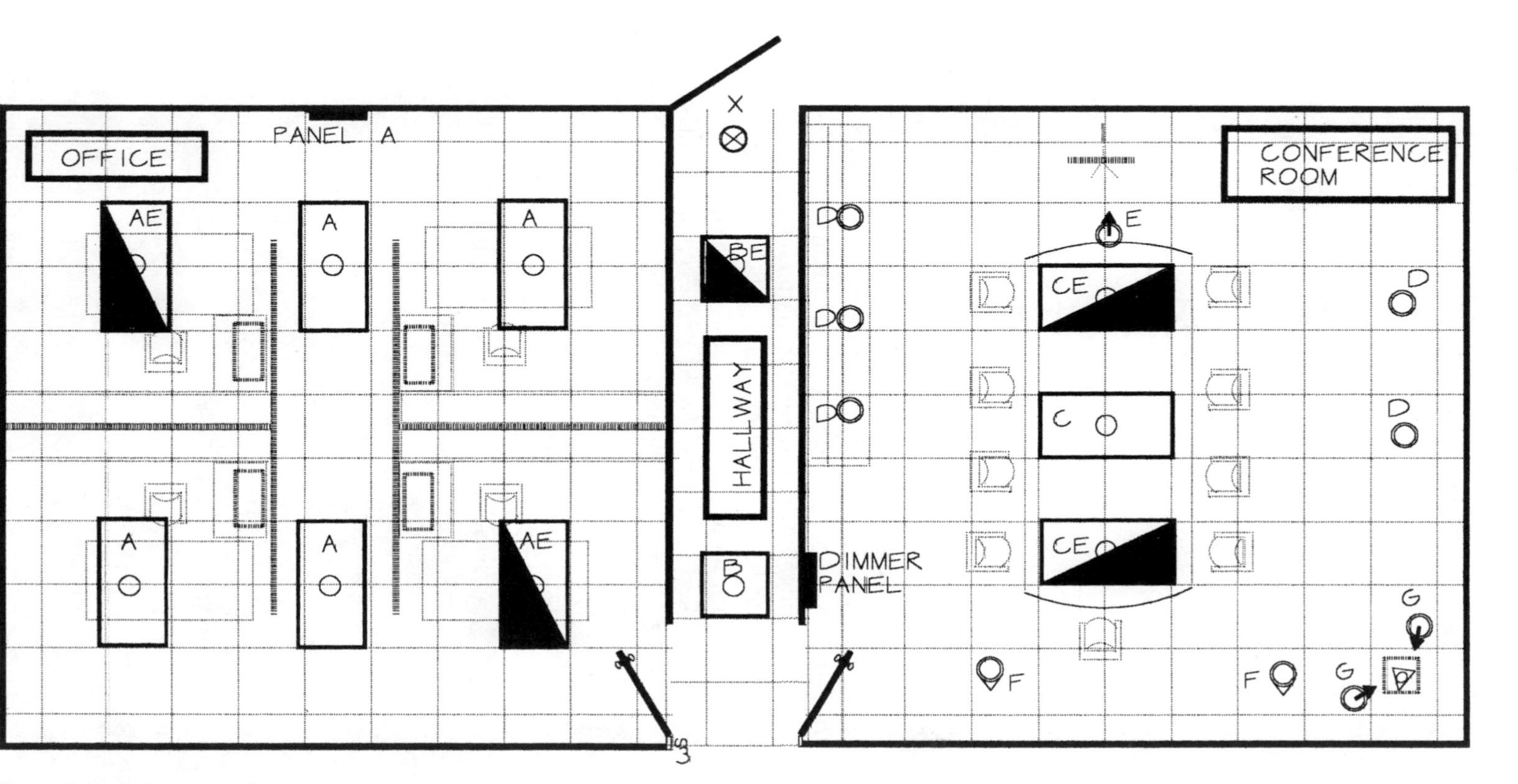

Figure 3.13. Bullmoose office addition lighting layout.

we need to think about the intent of the code, rather than the letter of the code, in order to do a good job of lighting for life safety. The code stresses that the paths of egress be illuminated. This means that, were you inside the space when the power went out, you would need enough light to safely exit the space without tripping over something.

Let's look at the office space. We have six luminaires spaced evenly in the ceiling. It would make sense to provide emergency lighting in the back of the room, farthest from the door, since the workers there have more obstacles in their path of egress. It also makes sense to put emergency lighting near the door, so the way out can be clearly seen. We can use two emergency luminaires and have the office space well covered. Now for the conference room, where all those foreign investors are watching a corporate presentation. We'll use a little overkill here, since it is important that the guests feel comfortable, even in the event of a power outage. We have three 2 ft × 4 ft luminaires above the conference table, so we will put battery backup in the two outboard ones. That will provide plenty of light. Out in the hallway, where everybody will be, once they leave their rooms, we need to provide both illumination and direction to the way out. We will provide battery backup in the luminaire closest to the door, and an illuminated exit sign with battery backup above the door leading to the outside. The exit sign will be illuminated by an LED (light emitting diode) source because of the low wattage and long lifetime of the LEDs. This should lead everybody out safely.

Now our layout for Bullmoose, Inc. is complete. Next, we need to identify the luminaires that we have used. To do this, we have assigned a luminaire type for each of the luminaires in our layout, and these are shown in the final layout of Figure 3.13.

Table 3.1. Luminaire schedule.

Type	*Manufacturer*	*Volts*	*Catalog no.*	*Lamps*	*Ballast*	*Comments*
A	Lithonia	120	2AVG332 ADP 120	3-F32T8/741	2-2F32 Elect	Office double sw
AE	Lithonia	120	2AVG332 ADP EL 120	3-F32T8/741	2-2F32 Elect	W/emerg batt pack
B	Lithonia	120	2SP2CF400A12 120	2-40WTT5	Electronic	Hallway
BE	Lithonia	120	2SP2CF40A12EL 120	2-40WTT5	Electronic	W/emerg batt pack
C	Lithonia	120	2PM433218L 120	3-F32T8/741	Elec dimming	Conf table
CE	Lithonia	120	2PM433218L EL 120	3-F32T8/741	Elec dimming	W/emerg batt pack
D	Lithonia	120	LGF 42TRT 75B 120 ADEZ	4-2WTRT	Elec dimming	Side table
E	Lithonia	120	DLVADJ MR'16 4ACT30 120	7-5WMR16 MF		Accent-easel
F	Lithonia	120	DLV WSH MR16 4AC 120	7-5WMR16 FL		Wallwash
G	Lithonia	120	DLVADJ MR'16 4ACT30 120	7-5WMR16 NS		Accent-bust
X	Lithonia	120	LE S W I R 120ELNSD	Red led		exit

Adapted with permission from Bullmoose, Inc.

Now we need to prepare a luminaire schedule, which will tabulate the luminaires that we have selected. The schedule should list each type of luminaire, and contain all pertinent information on the luminaires that we have chosen. Table 3.1 is a good format for the luminaires in the Bullmoose project.

All we have to do now is send the architect an invoice for our services, and then go out and party, right? Not quite. We still have to figure out how to make the design work, and put it all on a set of plans, so that a contractor can build it.

The next chapters will help us to do this. First though, let's try a few of those fun exercises to see what you have remembered from this chapter.

Exercises

1. What do you need to know about a space before you can begin lighting design?
2. What are the four basic types of lighting?
3. What does 'work plane' mean in ambient lighting?
4. What temperature lighting would normally be used for ambient lighting in a restaurant? Why?
5. If there were no constraints, which type of luminaire would you choose for low glare and even ambient light distribution?
6. What is a lighting manufacturer's representative, and why is he or she important to our design efforts?
7. What does it mean when the parallel and the perpendicular photometric curves of a luminaire are almost identical?
8. What is the efficacy of a CF40T5/741 lamp with an output of 3300 lumens?
9. Would a layout with a max/min ratio of 2.5:1 provide better light distribution than one with a max/min ratio of 1.5:1? Why?
10. How many footcandles at floor level are required for hallways?
11. What is the purpose of task lighting?
12. What does 'IC rated' mean for a recessed downlight?
13. What type of luminaire provides the majority of accent lighting in retail spaces?
14. How many footcandles do we need on the paths of egress during emergency conditions?
15. When would we use a centralized battery system for emergency lighting?
16. What are the two things we are trying to accomplish with emergency lighting?
17. Why do we specify an LED source in exit signs?

4 Powering and controlling the system

Now that we have selected the luminaires, and put together the layout for Bullmoose, Inc., we need to make the system function. To accomplish this, a number of decisions must be made, some of which could influence or negate our wonderful layout. The process of lighting design, as stated before, is an iterative process that offers the best compromise between the artist's vision and the real world.

Let's look now at some of the components involved in the power and control of lighting in general, and then we'll get back to the Bullmoose project.

Power

System voltage

One major concern is the voltage available to power the lighting system. Power companies in the US have standardized on three delivery voltages for buildings. These are 480Y/277 V three phase, 208Y/120 V three phase, and 240/120 V single or three phase. Larger buildings will typically have three-phase power available to drive the large heating and cooling loads, while smaller buildings often are served only with single-phase power.

Lighting loads are all single phase, so the lighting designer has the choice of 277 V or 120 V. Except in rare cases, incandescent lighting will only operate on 120 V, so if the building is served with 480Y/277 V, a step-down transformer must be used to power incandescent luminaires. We must then evaluate the economics of installing the transformer versus going to a ballasted light source, such as fluorescent or high intensity discharge (HID) lamps, which will operate on either 277 or 120 V.

Often, in a large project, 277 V is more economically attractive than 120 V, and to see how this works, let's consider how the system will be wired.

Wire sizing

Standard practice is to limit the size of the wiring serving lighting loads to No. 12 AWG, because anything larger than No. 12 is difficult to connect inside a luminaire, and labor costs increase dramatically (as do muttered insults by contractors) when larger wiring is used. No. 12 wire is rated for 25 A current load by code, and this limits the number of luminaires which can be connected to a circuit. To complicate matters further, the circuit must be protected by a 20 A molded case circuit breaker, which can only be loaded to 16 A by code. Each lighting circuit, therefore, is limited to 16 A load. A quick review of the basic electrical relationship $P = I \times E$ will reveal that for the allowable load of 16 A at 120 V, only nineteen 100 W luminaires can be connected on a circuit, while 16 A at 277 V will allow forty-four 100 W luminaires to be connected. If you're not up on electrical relationships, P is power in watts, I is current in amps, and E is voltage. Simply put, watts equals amps multiplied by volts.

It's easy to see that where the objective is to serve the maximum number of luminaires with the least number of circuits, 277 V would be the choice. This is applicable where it is desirable to serve large areas, such as open office spaces, by a single panelboard, which is limited to 42 breaker spaces by code.

Panelboards

Panelboards which serve lighting loads are standard circuit breaker panels, mounted either flush into, or on the surface of a wall. The panelboard assembly consists of a metallic box, called a *can*, which houses an interior section consisting of *vertical busses*, to distribute power, *circuit breakers* to serve the individual loads, and *termination blocks* for neutral and grounding connections. A three-phase panel contains three vertical busses, and a single-phase panel contains two busses.

The circuit breakers either snap or are bolted onto the busses to provide the power, or 'hot' connection for the lighting circuit. The circuit breakers are designed to automatically open the circuit in the event of overcurrent, and also serve as an on–off switch for the circuit. The final element of the panelboard is the *cover*, which contains a lockable door to keep the public out of the circuits, and which also trims out the opening made for the can.

Lighting loads may be directly controlled from the panelboard itself if a special switching duty circuit breaker, called a SWD breaker, is used. This is useful in large areas, such as in open plan office spaces, where it is desirable to switch large areas as a block load. In these cases, SWD breakers of either one, two, or three poles may be used to control the power from two or three panelboard busses simultaneously. For example, 132–100 W luminaires could be switched together using a three pole, 277 V, 20 A switching duty breaker, with 44 luminaires connected to each of the three panelboard busses.

It is always a good idea, when using a three-phase panelboard, to distribute the lighting loads as evenly as possible between the busses. This is called 'balancing the load'. Unbalanced loading can create large neutral currents and overheating within the panelboard. Figure 4.1 illustrates how the individual circuit breakers are connected to the busses in a three-phase panelboard.

Balancing the loads is an iterative operation that is most easily accomplished through the use of a panel schedule, such as the one shown in Figure 4.2.

Lighting loads are assigned to the busses sequentially, and then the total bus loads are calculated by adding the individual loads together. If the bus loads significantly differ from one another, the designer must reassign the loads until balance is obtained.

PANEL CONNECTION DIAGRAM

CKT NO	LOAD DESCRIPTION	LOAD (W)	BUSSES A B C	LOAD (W)	LOAD DESCRIPTION	CKT NO
1						2
3						4
5						6
7						8
9						10
11						12
13						14
15						16
17						18
19						20
21						22
23						24
25						26
27						28
29						30
31						32
33						34
35						36
37						38
39						40
41						42

Figure 4.1. Three-phase panelboard diagram.

PANEL: ______________________ VOLTAGE: ______________________

TYPE: ______________________ MAINS: ______________________

PANEL NOTE NO.	SIZE CONDUIT	WIRE			GRD.WIRE SIZE	DEVICE			BRANCH CIRCUIT			PHASE LOAD (VOLT-AMPS)			BRANCH CIRCUIT			DEVICE			GRID. WIRE SIZE	WIRE			SIZE CONDUIT	PANEL NOTE NO.
		NO.	SIZE	TYPE		AMP.TRIP	POLE	KAIC	DESIGNATION	VOLT-AMPS	NO.	ØA	ØB	ØC	NO.	VOLT-AMPS	DESIGNATION	KAIC	POLE	AMP.TRIP		TYPE	SIZE	NO.		
											1				2											
											3				4											
											5				6											
											7				8											
											9				10											
											11				12											
											13				14											
											15				16											
											17				18											
											19				20											
											21				22											
											23				24											
											25				26											
											27				28											
											29				30											
											31				32											
											33				34											
											35				36											
											37				38											
											39				40											
											41				42											

Figure 4.2. Three-phase panelboard schedule.

Load calculations

To calculate incandescent loads, we only need to add up the wattage of all the luminaires connected to the circuit. If amp loads are desired, total wattage divided by voltage (120 or 277) yields amps load. Ballasted loads are somewhat different in that a power loss associated with the ballast must be included in the total. For fluorescent fixtures with magnetic ballasts, this loss will be approximately 10% of the lamp wattage. For example, a fixture containing two 34 W F34T12 lamps would present a load of 110% of (2 × 34 W), or 75 W. Fluorescent fixtures operating on electronic ballasts, however, often present a load of *less* than the combined wattages of the included lamps. This is because the electronic ballast, which turns on and off at high frequency, is 'off' a lot of the time. Typically, a fixture containing two 32 W F32T8 lamps would present a load of only 56 W. High intensity discharge ballast losses vary with the type and wattage of lamp. It is always advisable to consult the manufacturer's information when calculating HID loads. In all cases, single-phase lighting amp loads can be found by dividing the total luminaire wattage by the voltage across the load ($I = P/E$).

Wiring and raceways

Because of circuit breaker restrictions, the amp load on a single 20 A circuit breaker cannot exceed 16 A. This is safely within the 25 A allowed by code for a No. 12 AWG copper conductor wiring system. Copper conductors should always be used for lighting wiring systems, because of the corrosion and temperature problems associated with aluminum conductors.

Wiring insulation, or the jacket covering the copper conductor, is typically either rubber, denoted by an 'R' in the wire type designation, or thermoplastic, denoted by a 'T'. Thermoplastic insulation is thinner than the rubber, so the overall cable diameter is less in a 'T' type cable. The 'T' type cable allows more cables to be pulled in a given size conduit than the 'R', and generally results in lower job cost. For this reason, type 'T' insulation is recommended for lighting jobs requiring conduit.

Some luminaires generate a considerable amount of heat, so the wiring insulation has to have the ability to withstand heat. This capacity is denoted by an 'H' or 'HH' in the wire type designation, such as 'THHN'. The NEC table 310-16 lists three categories of temperature ratings for wiring insulation: 60, 75, and 90°C. For almost all lighting applications, the 75°C wire is adequate. In rare cases, such as enclosed high power fiber optic sources, 90°C wiring insulation is required.

70–126 ARTICLE 310 — CONDUCTORS FOR GENERAL WIRING

Table 310-16. Allowable Ampacities of Insulated Conductors Rated 0 through 2000 Volts, 60°C through 90°C (140°F through 194°F) Not More than Three Current-Carrying Conductors in Raceway, Cable, or Earth (Directly Buried), Based on Ambient Temperature of 30°C (86°F)

Size	Temperature Rating of Conductor (See Table 310-13)						Size
	60°C (140°F)	75°C (167°F)	90°C (194°F)	60°C (140°F)	75°C (167°F)	90°C (194°F)	
AWG or kcmil	Types TW, UF	Types FEPW, RH, RHW, THHW, THW, THWN, XHHW, USE, ZW	Types TBS, SA, SIS, FEP, FEPB, MI, RHH, RHW-2, THHN, THHW, THW-2, THWN-2, USE-2, XHH, XHHW, XHHW-2, ZW-2	Types TW, UF	Types RH, RHW, THHW, THW, THWN, XHHW, USE	Types TBS, SA, SIS, THHN, THHW, THW-2, THWN-2, RHH, RHW-2, USE-2, XHH, XHHW, XHHW-2, ZW-2	AWG or kcmil
	COPPER			ALUMINUM OR COPPER-CLAD ALUMINUM			
18	—	—	14	—	—	—	—
16	—	—	18	—	—	—	—
14*	20	20	25	—	—	—	—
12*	25	25	30	20	20	25	12*
10*	30	35	40	25	30	35	10*
8	40	50	55	30	40	45	8
6	55	65	75	40	50	60	6
4	70	85	95	55	65	75	4
3	85	100	110	65	75	85	3
2	95	115	130	75	90	100	2
1	110	130	150	85	100	115	1
1/0	125	150	170	100	120	135	1/0
2/0	145	175	195	115	135	150	2/0
3/0	165	200	225	130	155	175	3/0
4/0	195	230	260	150	180	205	4/0
250	215	255	290	170	205	230	250
300	240	285	320	190	230	255	300
350	260	310	350	210	250	280	350
400	280	335	380	225	270	305	400
500	320	380	430	260	310	350	500
600	355	420	475	285	340	385	600
700	385	460	520	310	375	420	700
750	400	475	535	320	385	435	750
800	410	490	555	330	395	450	800
900	435	520	585	355	425	480	900
1000	455	545	615	375	445	500	1000
1250	495	590	665	405	485	545	1250
1500	520	625	705	435	520	585	1500
1750	545	650	735	455	545	615	1750
2000	560	665	750	470	560	630	2000

CORRECTION FACTORS

Ambient Temp. (°C)	For ambient temperatures other than 30°C (86°F), multiply the allowable ampacities shown above by the appropriate factor shown below.						Ambient Temp. (°F)
21–25	1.08	1.05	1.04	1.08	1.05	1.04	70–77
26–30	1.00	1.00	1.00	1.00	1.00	1.00	78–86
31–35	0.91	0.94	0.96	0.91	0.94	0.96	87–95
36–40	0.82	0.88	0.91	0.82	0.88	0.91	96–104
41–45	0.71	0.82	0.87	0.71	0.82	0.87	105–113
46–50	0.58	0.75	0.82	0.58	0.75	0.82	114–122
51–55	0.41	0.67	0.76	0.41	0.67	0.76	123–131
56–60	—	0.58	0.71	—	0.58	0.71	132–140
61–70	—	0.33	0.58	—	0.33	0.58	141–158
71–80	—	—	0.41	—	—	0.41	159–176

*See Section 240-3.

1999 Edition NATIONAL ELECTRICAL CODE

Figure 4.3. NFPA-70 (NEC) ampacity table (source: NFPA).

The building codes for commercial installations usually require lighting circuits to be pulled through conduit. If the pulls are difficult, or extremely long, an overall nylon jacket is available to protect the insulation from damage. This is denoted by an 'N' in the wire type designation. Finally, to protect against condensate, or moisture in the conduit, the cable should be rated for wet applications. This is denoted by a 'W' in the type designation.

Fortunately for lighting designers, almost all cable manufacturers make a type THHN/THWN cable, which is ideally suited for lighting applications. Note: residential installations, which do not require conduit, are normally wired with type NM (Non-Metallic sheathed) cable. Type NM is only rated for 60°C, and could cause problems with high temperature luminaires. Since we are primarily concerned with commercial installations, we will only use 75°C wiring pulled in conduit. Figure 4.3 is the NEC ampacity table, which lists current carrying capacity of type THHN/THWN wire in conduit.

The conduits through which lighting circuits are pulled are nothing more than pipes, and they are manufactured from several materials, each having a special purpose. Underground or in-slab conduit runs utilize PVC (polyvinyl chloride) conduit because of its resistance to corrosion. Conduit runs inside the structure are made in electrical metallic tubing (EMT) for mechanical protection, and also to provide electromagnetic shielding for sensitive equipment. In extremely harsh environments, such as in heavy industrial plants, circuits may be run in PVC-covered metal conduits.

Lighting circuits are usually run in hard conduit to a junction box in the immediate vicinity of the luminaires served. The wiring from the junction box to the individual luminaires is run in flexible metallic conduit (greenfield) for ease of connection. The code allows six feet of flexible cable to be used for this purpose. Figure 4.4 is a NEC table that specifies the maximum number of wires that may be pulled in a conduit.

Lighting controls

After the luminaires have all been selected and arranged within the space, it's time to consider how the system will be controlled to offer the occupants the most beneficial use of the space.

Switches

The most common form of lighting control is by the single pole, manually operated toggle switch. Switches are typically located

Table C1. Maximum Number of Conductors and Fixture Wires in Electrical Metallic Tubing
(Based on Table 1, Chapter 9)

CONDUCTORS											
	Conductor Size (AWG/ kcmil)	Trade Size (in.)									
Type		½	¾	1	1¼	1½	2	2½	3	3½	4
RH	14	6	10	16	28	39	64	112	169	221	282
	12	4	8	13	23	31	51	90	136	177	227
RHH, RHW, RHW,	14	4	7	11	20	27	46	80	120	157	201
	12	3	6	9	17	23	38	66	100	131	167
RH, RHH, RHW, RHW-2	10	2	5	8	13	18	30	53	81	105	135
	8	1	2	4	7	9	16	28	42	55	70
	6	1	1	3	5	8	13	22	34	44	56
	4	1	1	2	4	6	10	17	26	34	44
	3	1	1	1	4	5	9	15	23	30	38
	2	1	1	1	3	4	7	13	20	26	33
	1	0	1	1	1	3	5	9	13	17	22
	1/0	0	1	1	1	2	4	7	11	15	19
	2/0	0	1	1	1	2	4	6	10	13	17
	3/0	0	0	1	1	1	3	5	8	11	14
	4/0	0	0	1	1	1	3	5	7	9	12
	250	0	0	0	1	1	1	3	5	7	9
	300	0	0	0	1	1	1	3	5	6	8
	350	0	0	0	1	1	1	3	4	6	7
	400	0	0	0	1	1	1	2	4	5	7
	500	0	0	0	0	1	1	2	3	4	6
	600	0	0	0	0	1	1	1	3	4	5
	700	0	0	0	0	0	1	1	2	3	4
	750	0	0	0	0	0	1	1	2	3	4
	800	0	0	0	0	0	1	1	2	3	4
	900	0	0	0	0	0	1	1	1	3	3
	1000	0	0	0	0	0	1	1	1	2	3
	1250	0	0	0	0	0	0	1	1	1	2
	1500	0	0	0	0	0	0	1	1	1	1
	1750	0	0	0	0	0	0	1	1	1	1
	2000	0	0	0	0	0	0	1	1	1	1
TW	14	8	15	25	43	58	96	168	254	332	424
	12	6	11	19	33	45	74	129	195	255	326
	10	5	8	14	24	33	55	96	145	190	243
	8	2	5	8	13	18	30	53	81	105	135
RHH*, RHW*, RHW-2*, THHW, THW THW-2,	14	6	10	16	28	39	64	112	169	221	282
RHH*, RHW*, RHW-2*, THHW, THW	12	4	8	13	23	31	51	90	136	177	227
	10	3	6	10	18	24	40	70	106	138	177
RHH*, RHW*, RHW-2*, THHW, THW, THW-2	8	1	4	6	10	14	24	42	63	83	106

*Types RHH, RHW, and RHW-2 without outer covering.

Table C1. Continued

CONDUCTORS											
	Conductor Size (AWG/ kcmil)	Trade Size (in.)									
Type		½	¾	1	1¼	1½	2	2½	3	3½	4
RHH*, RHW*, RHW-2*, TW, THW, THHW, THW-2	6	1	3	4	8	11	18	32	48	63	81
	4	1	1	3	6	8	13	24	36	47	60
	3	1	1	3	5	7	12	20	31	40	52
	2	1	1	2	4	6	10	17	26	34	44
	1	1	1	1	3	4	7	12	18	24	31
	1/0	0	1	1	2	3	6	10	16	20	26
	2/0	0	1	1	1	3	5	9	13	17	22
	3/0	0	1	1	1	2	4	7	11	15	19
	4/0	0	0	1	1	1	3	6	9	12	16
	250	0	0	1	1	1	3	5	7	10	13
	300	0	0	1	1	1	2	4	6	8	11
	350	0	0	0	1	1	1	4	6	7	10
	400	0	0	0	1	1	1	3	5	7	9
	500	0	0	0	1	1	1	3	4	6	7
	600	0	0	0	1	1	1	2	3	4	6
	700	0	0	0	0	1	1	1	3	4	5
	750	0	0	0	0	1	1	1	3	4	5
	800	0	0	0	0	1	1	1	3	3	5
	900	0	0	0	0	0	1	1	2	3	4
	1000	0	0	0	0	0	1	1	2	3	4
	1250	0	0	0	0	0	1	1	1	2	3
	1500	0	0	0	0	0	1	1	1	1	2
	1750	0	0	0	0	0	0	1	1	1	2
	2000	0	0	0	0	0	0	1	1	1	1
THHN, THWN, THWN-2	14	12	22	35	61	84	138	241	364	476	608
	12	9	16	26	45	61	101	176	266	347	443
	10	5	10	16	28	38	63	111	167	219	279
	8	3	6	9	16	22	36	64	96	126	161
	6	2	4	7	12	16	26	46	69	91	116
	4	1	2	4	7	10	16	28	43	56	71
	3	1	1	3	6	8	13	24	36	47	60
	2	1	1	3	5	7	11	20	30	40	51
	1	1	1	1	4	5	8	15	22	29	37
	1/0	1	1	1	3	4	7	12	19	25	32
	2/0	0	1	1	2	3	6	10	16	20	26
	3/0	0	1	1	1	3	5	8	13	17	22
	4/0	0	1	1	1	2	4	7	11	14	18
	250	0	0	1	1	1	3	6	9	11	15
	300	0	0	1	1	1	3	5	7	10	13
	350	0	0	1	1	1	2	4	6	9	11
	400	0	0	0	1	1	1	4	6	8	10
	500	0	0	0	1	1	1	3	5	6	8
	600	0	0	0	1	1	1	2	4	5	7
	700	0	0	0	1	1	1	2	3	4	6
	750	0	0	0	0	1	1	1	3	4	5
	800	0	0	0	0	1	1	1	3	4	5
	900	0	0	0	0	1	1	1	3	3	4
	1000	0	0	0	0	1	1	1	2	3	4
FEP, FEPB, PFA, PFAH, TFE	14	12	21	34	60	81	134	234	354	462	590
	12	9	15	25	43	59	98	171	258	337	430
	10	6	11	18	31	42	70	122	185	241	309
	8	3	6	10	18	24	40	70	106	138	177
	6	2	4	7	12	17	28	50	75	98	126
	4	1	3	5	9	12	20	35	53	69	88
	3	1	2	4	7	10	16	29	44	57	73
	2	1	1	3	6	8	13	24	36	47	60
PFA, PFAH, TFE	1	1	1	2	4	6	9	16	25	33	42
PFA, PFAH, TFE, Z	1/0	1	1	1	3	5	8	14	21	27	35
	2/0	0	1	1	3	4	6	11	17	22	29
	3/0	0	1	1	2	3	5	9	14	18	24
	4/0	0	1	1	1	2	4	8	11	15	19
Z	14	14	25	41	72	98	161	282	426	556	711
	12	10	18	29	51	69	114	200	302	394	504
	10	6	11	18	31	42	70	122	185	241	309
	8	4	7	11	20	27	44	77	117	153	195
	6	3	5	8	14	19	31	54	82	107	137
	4	1	3	5	9	13	21	37	56	74	94
	3	1	2	4	7	9	15	27	41	54	69
	2	1	1	3	6	8	13	22	34	45	57
	1	1	1	2	4	6	10	18	28	36	46

*Types RHH, RHW, and RHW-2 without outer covering.

NATIONAL ELECTRICAL CODE 1999 Edition

Figure 4.4. NFPA-79 (NEC) conduit allowable fill table (source: NFPA).

within the space to be lighted, usually on the lock side of the entrance door, mounted 48 in. (122 cm) above the floor. If there is more than one door leading into a space, three-way, and four-way toggle switches are available to allow switching from any door location. Figure 4.5 contains connection diagrams for toggle switching.

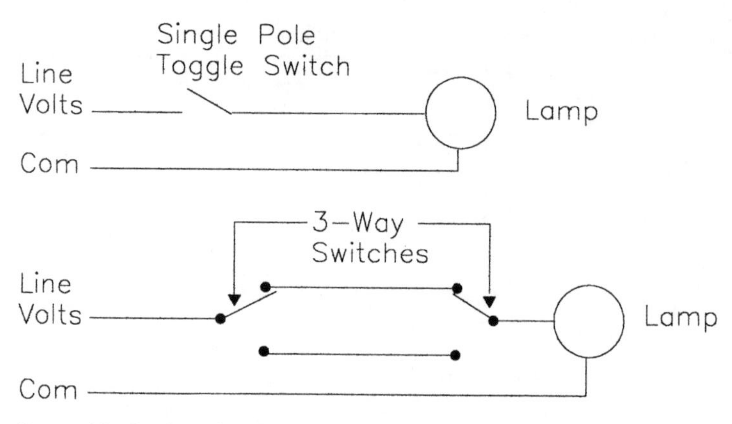

Figure 4.5. Single and three-way switching.

Switches are available for both 277 V and for 120 V, and are available in either 15 or 20 A ratings. The 20 A switches have slightly heavier contacts, and their use is recommended. The total luminaire amperage on a circuit cannot exceed the switch rating, so if you push the circuit to its full 16 A capacity, you *must* use 20 A rated switches.

Toggle switches are either on or off, and have no light dimming capability. When used in conjunction with multi-ballasted fluorescent fixtures, however, they can offer some lighting level control. Figure 4.6 is a connection diagram of a double-switched, two-ballast, three-lamp fluorescent luminaire that provides three light levels: one, two, or all three lamps can be illuminated by the use of the two toggle switches.

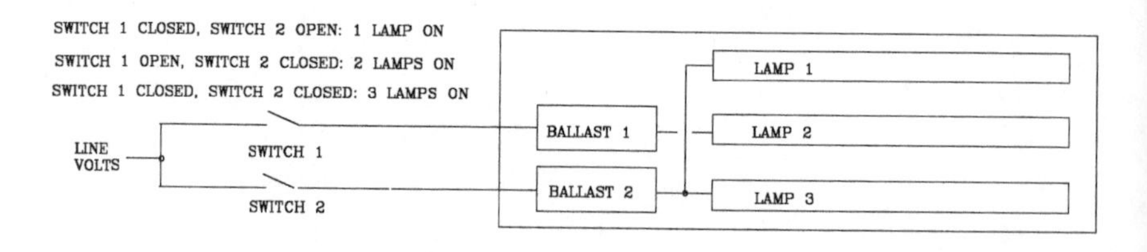

DOUBLE SWITCHING OF A 3–LAMP LUMINAIRE

Figure 4.6. Double-switched three-lamp luminaire.

Dimmers

When continuous dimming of the light level is desired, such as in churches, restaurants, theaters, etc., specialized dimming equipment, tailored to the luminaires to be dimmed must be used. Since dimmers are much more expensive than on–off switching, economics will often dictate the type and amount of dimming used.

Dimmers do as their name implies: they dim, or reduce the light output of the luminaires attached to them. The type of dimmer most widely in use today is the simple rheostat, or variable resistor, which is used to dim incandescent loads. Rheostats directly vary the amount of current flowing through the incandescent filament, and by doing so, they vary the light output of the lamp. Rheostats are line voltage, and come in various sizes, which indicate the maximum amount of wattage that they are capable of controlling. Common rheostats are available up to 2000 W, which is the maximum load that can be put on a 20 A circuit at 120 V.

Dimming a fluorescent luminaire requires a special dimming ballast. Magnetic fluorescent dimming ballasts operate from line voltage, and are normally used in conjunction with a single circuit, wall-mounted dimming control. Electronic fluorescent dimming ballasts are controlled by a separate low voltage dimming circuit, usually 24 V. This arrangement is usable with locally mounted controls, and is also suitable for zone control of block loads. This is particularly applicable to computerized dimming control that offers multiple preset 'scenes' such as that used in churches and performing arts theaters. In operation, this system allows the user to set the level of several groups of luminaires through the use of slide or rotary switches, and then 'save' the settings as a scene. To recall a scene, the user needs only to push a single button, and the saved settings are automatically retrieved by the dimming system.

If groups of high wattage luminaires are to be dimmed, such as in large churches or theaters, specialized dimming systems capable of handling large power loads are used. These systems are designed, building block style, from rack-mounted components specifically selected for the purpose.

Contactors

When it is desired to control a larger group of luminaires than a single-switched circuit can handle, and multiple switches are undesirable, as is common in large office spaces, lighting contactors are used. Contactors are essentially solenoids, or relays, in which a low power coil actuates, or 'pulls in' a set of heavy contacts that applies power to the lighting load. The coil itself can be activated by a single manual control device such as a wall-mounted toggle switch, or it may be activated by automatic controllers, such as photocells, occupancy sensors, timers, or building management computer systems. Contactors typically have from 2 to 12 sets of contacts, rated at 30 A each. Larger contactor ratings are available for specialized purposes. Each set of contacts can be used to power a separate lighting circuit.

Once activated, contactors are either mechanically or electrically held in the closed position. In the mechanically held contactor, the mechanism closes when the coil is activated, and it is held closed by a mechanical 'latch' until a second coil is activated, which 'unlatches' the mechanism. This requires only a momentary application of coil power through the switching device. The electrically held contactor requires that coil power be continuously applied while the contactor is closed. Removal of coil power allows the spring-loaded contactor to open. The electrically held contactor is less expensive than the mechanically held, but it can produce an undesirable buzzing sound while the coil is energized.

Photocells

Photocells are basically devices used to turn the lights on when it is dark, and to turn them off when daylight is present. A photocell consists of a photovoltaic cell connected to a current-sensitive relay. When light strikes the cell, an electric current proportional to the intensity of the light is generated. When that current is sufficient to activate the relay coil, normally closed relay contacts open, and interrupt the power to a lighting circuit or contactor. When low light is present, the contacts close, and apply power to the lighting circuit. The contacts are normally rated for only 5 or 10 A, so the photocell can directly control only a small number of luminaires. When the lighting circuit load exceeds the rating of the photocell contacts, a contactor must be used in conjunction with the photocell.

Some luminaires designed for outside use offer an attached photocell as an option for controlling that luminaire only. Photocells are normally used to control exterior lighting by turning the luminaires on at dusk and off at dawn. Common practice is to mount the photocell on the north side of the building. For obvious reasons, photocells should not be used in extremely dirty environments.

Timers

A lighting timer is a device that turns luminaires on or off at preset times. Timers can be simple electro-mechanical devices with manual stops that provide on–off switching, or sophisticated devices programmed for sunrise and sunset times throughout the year. They can be built into the program of a building management system so that the lights come on at the beginning of the workday, and go off at quitting time. There is a timer available to provide any desired schedule for any facility. The only limiting factor in providing timer control is economics.

Occupancy sensors

Occupancy sensors are devices used to turn on lights when people are in the space, and to turn them off when the space is vacant. They are widely used in office spaces to conserve energy, and in sensitive areas for security.

Three types of occupancy sensors are available: *passive infrared*, or PIR, *ultrasonic*, and *dual technology*, which is a combination of the two. All of the sensors contain timers that delay the action of the switch for a preset time.

The PIR sensor detects the difference in infrared energy between a human body and the surrounding space. When sufficient infrared energy is detected, it closes a set of normally open contacts to turn on the lights. The PIR sensor controls an area within its line of sight, and within its coverage zone. False 'ON' can occur when someone passes a doorway within sight of the sensor, and a false 'OFF' can occur if the room occupant is obscured by a piece of furniture.

PIR sensors are best suited for:

1. enclosed open offices;
2. warehouses;
3. hallways;
4. areas with high air flow;
5. high ceiling areas.

They are poorly suited for:

1. restrooms;
2. storage areas with shelving;
3. offices with partitions.

The ultrasonic sensor uses the Doppler principle to sense motion within the space: it bounces ultrasonic sound waves off objects in the space, and measures the time that it takes the waves to return. Movement by a person in the space disrupts the return time, and the sensor reacts by closing a set of contacts. False 'ON' can occur if there is high air flow within the space. False 'OFF' can occur if the occupant of the space is inactive for longer than the delay period.

Ultrasonic sensors work best in:

1. open plan office spaces;
2. conference rooms;
3. restrooms;
4. large areas.

They work poorly in:

1. spaces with high air flow;
2. spaces with high ceilings;
3. unenclosed areas.

The dual technology sensor utilizes both PIR and ultrasonic sensing, and will not react unless *both* sensors detect occupancy, or the lack thereof. This eliminates false 'ON' and false 'OFF' in most instances.

Dual technology sensors perform best in:

1. classrooms;
2. office spaces with partitions;
3. areas with high ceilings;
4. large open areas.

The only drawback to dual technology sensors is their relatively high cost.

Occupancy sensors range in size from those which replace toggle switches for single office applications, to those which are designed to cover warehouses.

Light-sensitive controls

Light-sensitive controls are sensors that measure the ambient daylight in a space, and then control the artificial lighting to maintain a preset lighting level. These sensors contain an internal photo conductive cell that measures light levels, and after a preset delay, they either cut off luminaires, or dim them to a preset level. These controls have been used with limited success, largely in areas of the country that have minimal cloud cover. Obviously, the timing of cloud passage cannot be predicted, so the setting of the delay can present problems in areas with a high incidence of cloud cover. These sensors contain internal switches for on–off control, and electronic circuitry for electronic ballast control.

Example

Now that we're all experts in powering and controlling lighting systems, let's take another look an Bullmoose Inc.'s office and conference room addition, and see how we would make the system work. The maintenance supervisor for Bullmoose has told us that the voltage available in the building is 208Y120 V, three phase, and the nearest panelboard to our area is several hundred feet away.

Instead of pulling all of our lighting circuits for that distance, it makes more sense to install a small new 208Y120 V panelboard in our area, and only pull one feeder circuit from the existing panelboard. We know that there will be other circuits needed in the area for receptacles, equipment, etc., so the cost of the new panelboard is justified. We will flush mount the new panelboard in the office space, out of sight from visitors. Now let's wire up the lighting.

Let's look at the office first. We can see from Figure 3.5 that we have chosen six direct/indirect luminaires, each with three 32 W lamps, for a total load of 576 W, assuming no reduction for our electronic ballasts. At 120 V, our amp load is 576/120, or about 5 A. This is well within the capacity of one 20 A circuit, which you will remember, is good for 16 A. This is also within the capacity of a single 20 A switch, so we could control all the lights in the office with one switch. For just a little extra money, we can put in another switch, and allow control of the lighting level with the switching scheme shown in Figure 4.6. Let's do that. That will give the office workers some control over the brightness in their room.

Standard practice is to put the switch on the lock side of the door, so that a person entering the room may reach in and turn on the lights without searching for the switch. We will assign one circuit to the office lighting by itself, even though we have spare capacity, so if the circuit breaker serving the office is tripped off, only the office area will be affected.

Now what happens when everybody in the office goes to lunch for an hour? Do you think that one of the workers will remember to turn the lights off to save energy? Don't count on it. With just a little more expense, we can install an occupancy sensor to do the job every time. We will need a dual technology sensor, since the office has partitions in it, and the sensor will work best if it is mounted in the ceiling in the center of the room so that it can 'see' into all the cubicles. The sensor should be wired ahead of, and in series with the wall switches, so that it is only active when a switch is in the 'on' position.

That done, let's have a look at the conference room, which will be a bit trickier. In reviewing the requirements for that room we see that we will need full light over the conference table and credenza when meetings are taking place, but only minimal light there when a presentation is in progress. Our accent light for the easel should be on full when a presenter is showing charts, but off when a video presentation is in progress. Our accent lighting for the art features, such as the portraits and the bust of the Commodore should be on full for meetings, but subdued for presentations. Upon closer examination, several scenes begin to arise from these requirements.

Our architect has told us that Bullmoose Inc. doesn't mind spending a little money on the conference room lighting, so we will use a small computerized dimming system that has the capacity for, say, four scenes. We will start with each of our lighting functions separately circuited, each adjustable from a slider switch. The conference table and credenza fluorescent ballasts will require a low voltage signal, and the MR16 accent lights will be dimmed directly from line voltage. The load for the 3–3 lamp conference table luminaires will be 3 × 3 × 32 W, or 288 W. The credenza lighting and east wall load will be 5 × 42 W, or 210 W. The accent luminaires all use 75 W MR16 lamps, so the easel light load will be 75 W, and the artwork lighting will be 4 × 75 W, or 300 W. Obviously, none of these loads, or even all of these loads combined, will exceed the 2000 W limit of one circuit. So instead of using four power circuits for the conference room, we will use only two circuits, one for the fluorescent dimmers, and one for the MR16 dimmers.

Now, would there be any benefit to using occupancy sensors for the conference room? Probably not, because the same company ferret who is assigned to make sure that the coffee and pastries arrive on time, will also be responsible for turning the lights on and off.

That takes care of circuiting the office and conference room. What about the hallway – how will we handle the switching for that, since the hallway can be entered from either end? This is where we use three-way switching, as shown in Figure 4.5. We will use one circuit for the hallway, and we have two luminaires with two 40 W lamps each, or a total of 160 W.

Now what about power for the battery chargers for the emergency and exit lighting? We will want to put those on their own circuit, and label the circuit 'EMERGENCY LIGHTS – DO NOT TURN OFF' by placing a sticker inside the panelboard beside the circuit breaker serving this load. Each charger requires about 50 W, and we have a total of five chargers, so the load is 250 W.

Now let's take a look at the loads that we have assigned to circuit breakers:

- 720 W office lighting
- 288 W + 210 W = 498 W conference room fluorescent lighting

- 75 W + 300 W = 375 W conference room MR16 accent lighting
- 160 W hallway lighting
- 250 W emergency lighting battery chargers

Let's use the panel schedule shown in Figure 4.2 to assign these loads. If we put the 720 W office lighting load on circuit 1, which is on bus A, we find that if we use circuit 2, also bus A, we will have a heavy imbalance on bus A. So we will skip circuit 2, and assign the conference room fluorescent lighting to circuit 3, which is on bus B. We can assign the hallway lighting to circuit 4, also on bus B, and have a total of 754 W on bus B. We assign the 375 W conference room MR16 accent lighting load to circuit 5, on bus C, and the 250 W emergency battery charger load to circuit 6, which gives us a total load of 625 W load on bus C. This is not an ideal balance of loads, but it is as close as we can get with the loads that we have. This arrangement is shown in Figure 4.7. You will also see in Figure 4.7 that the panel schedule has space to define the circuit breaker and the circuit serving each load. As planned, we will use single pole circuit breakers and no. 12 AWG wire run in 0.5 in. conduit.

PANEL: A (SEE NOTE 3) TYPE:
VOLTAGE: 120/208 3P 4W MAINS: 100A MAIN LUGS

PANEL NOTE NO.	SIZE CONDUIT	POWER CABLE				DEVICE			BRANCH CIRCUIT			PHASE LOAD (KVA)				BRANCH CIRCUIT		DEVICE				POWER CABLE				
		NO.RUNS	NEUT	PHASE	GRD.	AMP.TRIP	POLE	KAIC	DESIGNATION	AMPS	NO.	φA	φB	φC	NO.	AMPS	DESIGNATION	KAIC	POLE	AMP.TRIP	GRD.	PHASE	NEUT	NO.RUNS	SIZE CONDUIT	PANEL NOTE NO.
	1/2	1	12	12	12	20	1	10	OFFICE LIGHTS		1	720			2		SPACE									
	1/2	1	12	12	12	20	1	10	DIMMER PANEL		3		658		4		HALLWAY	10	1	20	12	12	12	1	1/2	
	1/2	1	12	12	12	20	1	10	DIMMER PANEL		5			625	6		EXIT + EMERG	10	1	20	12	12	12	1	1/2	
											7				8											
											9				10											
											11				12											
											13				14											
											15				16											
											17				18											
											19				20											
											21				22											
											23				24											
											25				26											
											27				28											
											29				30											
											31				32											
											33				34											
											35				36											
											37				38											
											39				40											
											41				42											

Figure 4.7. Bullmoose office addition panel schedule.

We've now put together a luminaire layout, a luminaire schedule, and a panel schedule for our space. What next? We'll find out, just as soon as we get these pesky exercises out of the way.

Exercises

1. What are the two voltages in the US of concern in lighting design?
2. What size wire is normally used for lighting circuits?
3. What is the allowable amperage load on a 20 A molded case circuit breaker?
4. What is the maximum number of circuit breaker spaces that can be built into a panelboard?
5. How many amps of current would be drawn by 12–100 W lamps connected to a 120 V circuit? A 277 V circuit?
6. What are the five components of a panelboard?
7. What are the functions of a circuit breaker in a lighting circuit?
8. What is meant by 'balancing the load' in a three-phase panelboard?
9. What percent of fluorescent lamp wattage do most magnetic ballasts consume as power loss?
10. What does the 'T' in the wiring designation THWN mean? The 'W'?
11. What does 'EMT' mean when referring to conduit?
12. Can a single ballast fluorescent luminaire be double switched? Why?
13. When an incandescent lamp is dimmed, does its color temperature increase or decrease? Why?
14. What is a 'scene' as applied to computerized fluorescent dimming?
15. What are the two methods for holding lighting contactors closed?
16. What is the most common use for the photocell?
17. How many types of occupancy sensors are there? What are they?
18. What type of occupancy sensor would you use in a school classroom?
19. Would you use light-sensitive control of ambient lighting in an area with a high incidence of cloud cover? Why?

5 The contract documents

OK, we have created a design for the Bullmoose, Inc. office addition using catalogued lamps and luminaires, and we have selected controls and circuits to make it work. Where do we go from here?

Now we have to pull all of these elements together, and create a set of contract documents that an electrical contractor can use to bid on, and then use to build the system. The designer's part of the contract documents usually consists of a set of drawings, which describe the system graphically, and the technical specifications, which describe the components that the designer wants the contractor to use. Terms of contract, bidding procedures, etc., are usually prepared by the lead professional, which in our case is the architect.

The contractor will first use these documents to prepare a bid for constructing the project. His bid will include his cost for the components shown on the drawings, plus his cost for labor to construct the project, plus any overhead and materials markup that he chooses to include. If the contractor is the successful bidder, he will then use the documents to build the system.

If you as a designer have left a required component off of the plans, or out of the specifications, it's a sure bet that the contractor will find the oversight. The same holds true for components that are sized incorrectly, and for designs that fail to meet the applicable codes. During construction, the contractor will request a change order to correct the deficiencies. Change orders invariably mean an increase in contract price, and they tend to make owners and architects very unhappy – even to the point of requiring the delinquent designer to pay for the changes out of his own pocket. You can see, then, how imperative it is for contract documents to be complete and accurate.

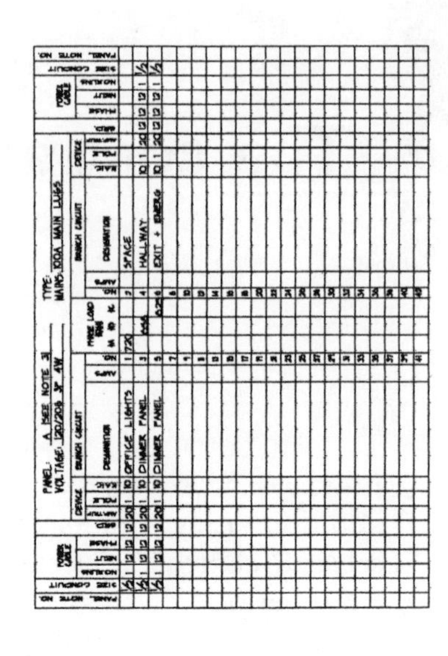

PANEL: A (SEE NOTE 3) VOLTAGE: 120/208 3Φ 4W TYPE: MAINS: 100A MAIN LUGS

PANEL NOTE NO.	SIZE CONDUIT	NO. RUNS	NEUT	PHASE	GRD.	AMP/TRIP	POLE	KAIC	DESIGNATION	AMPS	NO.	A	B	C	NO.	AMPS	DESIGNATION	KAIC	POLE	AMP/TRIP	GRD.	PHASE	NEUT	NO. RUNS	SIZE CONDUIT	PANEL NOTE NO.
	1/2	1	12	12	12	20	1	10	OFFICE LIGHTS		1	1720			2		SPACE									
	1/2	1	12	12	12	20	1	10	DIMMER PANEL		3		056		4		HALLWAY	10	1	20	12	12	12	1	1/2	
	1/2	1	12	12	12	20	1	10	DIMMER PANEL		5			025	6		EXIT + EMERG	10	1	20	12	12	12	1	1/2	
											7				8											
											9				10											
											11				12											
											13				14											
											15				16											
											17				18											
											19				20											
											21				22											
											23				24											
											25				26											
											27				28											
											29				30											
											31				32											
											33				34											
											35				36											
											37				38											
											39				40											
											41				42											

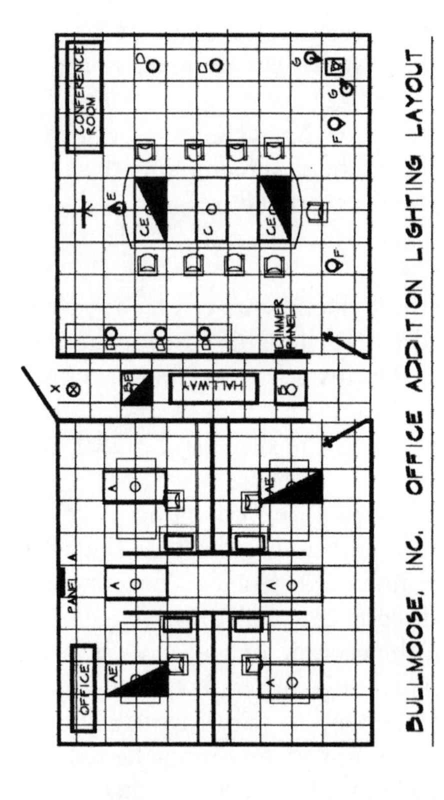

LUMINAIRE SCHEDULE

TYPE	MANUFACTURER	VOLTS	CATALOG NO.	LAMPS	BALLAST	COMMENTS
A	LITHONIA	120	2AV6332 ACP 120	3-F32T8/741	2-2F32 ELECT	OFFICE DOUBLE SW
AE	LITHONIA	120	2AV6332 ACP EL 120	3-F32T8/741	2-2F32 ELECT	W/ EMERG. BATT PACK
B	LITHONIA	120	2SP2CF40A12 120	2-40WTT5	ELECTRONIC	HALLWAY
BE	LITHONIA	120	2SP2CF40A12EL 120	2-40WTT5	ELECTRONIC	W/ EMERG. BATT PACK
C	LITHONIA	120	2TMA332I8L 120	3-F32T8/741	ELEC DIMMING	CONF TABLE
CE	LITHONIA	120	2TMA332I8L EL 120	3-F32T8/741	ELEC DIMMING	W/ EMERG. BATT PACK
D	LITHONIA	120	L6F 42TRT 730 120 ADEZ	42WTRT	ELEC DIMMING	CREDENZA
E	LITHONIA	120	OLVAOJ MR16 4AGT30 120	75WMR16 NF		ACCENT - EASEL
F	LITHONIA	120	OLY WSH MR16 4AC 120	75WMR16FL		WALLWASH
G	LITHONIA	120	OLVAOJ MR16 4AGT30 120	75WMR16 NS		ACCENT - BUST
X	LITHONIA	120	LE 2 W 1 R 120ELN50	RED LED		EXIT

FIG 5-1 PLAN SHEET START

BULLMOOSE INC. NEW OFFICE ADDITION
LIGHTING DETAILS

Figure 5.1. Bullmoose plan sheet start.

The project plans

So far, we've been pretty concise with our Bullmoose, Inc. design: life safety codes have been met; luminaires have been selected to meet our design requirements; and circuiting has all been well within NEC requirements. So let's put our luminaire schedule and panel schedule on a single sheet to begin our plan set. This is shown in Figure 5.1.

In order to show graphically what we intend the contractor to do, we need to develop a legend of symbols that have specific meaning on this drawing set. Standardized drawing symbols are published in many architectural reference books, such as *Architectural Graphic Standards* published by John Wiley and Sons, and other texts, all available from the IES publications division. Many computerized drawing programs also contain libraries of standard symbols. For our project, we will use the legend shown in Figure 5.2

Now we need to show the wiring on the plan view, and identify all the luminaires. This is shown in Figure 5.3.

All the wiring is identified in the panel schedule, with the exception of the extra conductor needed in the run between the two

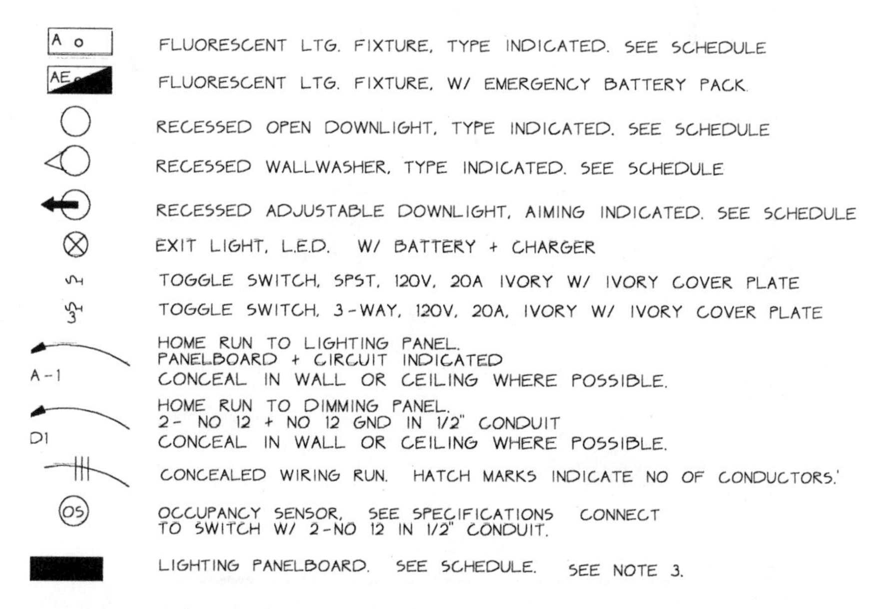

Figure 5.2. Legend for Bullmoose office addition plan.

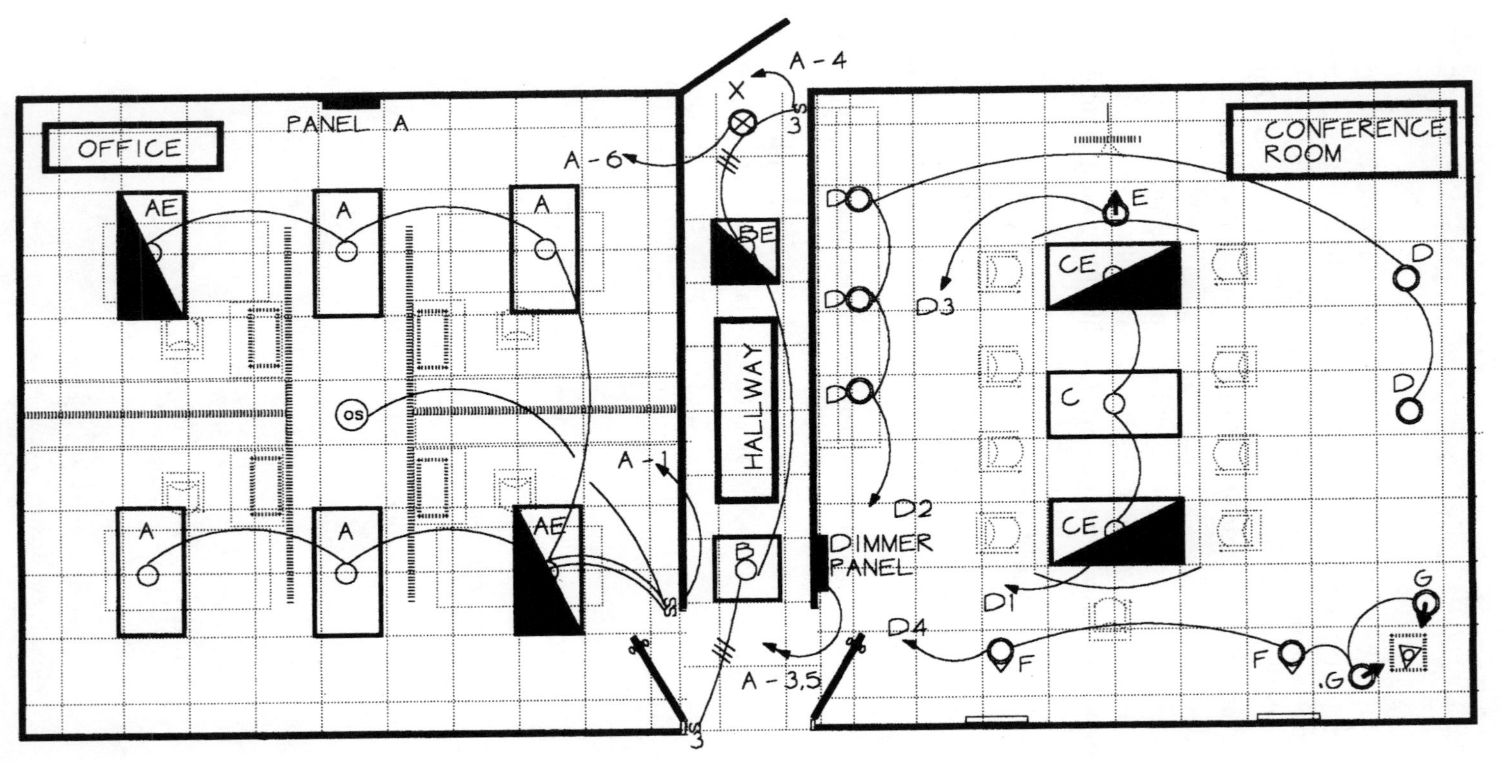

Figure 5.3. Wiring for Bullmoose office addition plan.

hallway three-way switches, which, if you recall from Figure 4.3, runs from switch to switch. We will indicate this extra conductor by hatching the switch leg of the hallway lighting, as shown in Figure 5.3.

The circuit breaker in the existing building panelboard, the feeder run from that panelboard to our new panelboard A, and new panelboard A should be provided by the electrical contractor, if he is different from the lighting contractor.

All the components and wiring are identified now, so our plans are almost complete. All we need to add are some clarifying notes and details to make sure that our intent is understood by the contractor. We need to make sure that the contractor installs the system in compliance with all applicable codes, so we'll make this note 1. Wiring is usually included in the electrical (rather than the lighting) specifications, but it's a good idea to specify the lighting wiring on the lighting plans as well. This will be note 2. We want to clarify responsibility for the circuit breaker in the existing building panel, the feeder to the new panel, and the new panel itself, so we will make this note 3. The note will indicate that the electrical contractor is to coordinate with the owner to identify the existing panel from which our new panel will be fed. Note 3 will also make reference to the electrical specification section which covers circuit breakers and panelboards. Note 4 will require that the contractor provide the exact luminaires that we show in the luminaire schedule. This is necessary, because this is where a lot of lighting designs go awry: most contractors have long-standing relationships with suppliers who only carry specific brands of equipment. If our contractor buys from a supplier who does not carry the luminaires that we have specified, he will want to substitute luminaires that his supplier does carry. The contractor is not aware of the design criteria, and he may ask to substitute a luminaire that is unsuitable for the job. Since the contractor's bid is final, it is incumbent upon the designer to make sure, before the bids, that he has the opportunity to accept or reject a contractor's request for substitutions. In our case, since we have worked closely with only one manufacturer's representative and products, it is safer to not allow any substitutions. A substitution clause will be included in note 4.

(A note on substitutions is appropriate here. If your project is funded by public money, such as schools or government buildings, it is mandated by law in most locations that you allow at least three manufacturer's products to be bid, in order to obtain the best competitive price. If the project is privately funded, as is the case with our Bullmoose, Inc. project, the designer may require that the project be built as shown on the plans and specifications, with no substitutions.)

NOTES

1. ALL WIRING TO BE PER N.E.C. AND THE LOCAL COUNTY AND STATE ELECTRICAL CODES
2. ALL WIRING TO BE 75 DEG C, TYPE THHN/THWN INSULATION COPPER CONDUCTOR, UNLESS NOTED OTHERWISE (U.N.O.)
3 ELECTRICAL CONTRACTOR WILL PROVIDE NEW PANEL A, NEW FEEDER CIRCUIT AND NEW CIRCUIT BREAKER IN OWNER DESIGNATED PANEL UNDER SECTION 16402. LIGHTING CONTRACTOR TO PROVIDE LIGHTING BRANCH CIRCUITS.
4 LUMINAIRES ARE TO BE PROVIDED AS SHOWN IN THE LUMINAIRE SCHEDULE. NO SUBSTITUTIONS WILL BE ACCEPTED.
5. LUMINAIRES ARE TO BE PROVIDED COMPLETE WITH ALL REQUIRED TRIM, HANGERS, AND OTHER APPURTENANCES NECESSARY FOR THEIR INSTALLATION.

Figure 5.4. Notes for Bullmoose office addition plan.

Note 5 is a general CYA (Cover Your A—-) note to prevent change order requests for things like trim, luminaire clips, lenses, etc., which the contractor should have known to include in his bid.

The completed notes are shown in Figure 5.4.

We will add the notes to the drawing. Now what about details? About the only thing in our design which isn't standard construction practice is the double switching of the office luminaires, so we'll add the detail shown in Figure 4.6, and that will complete the plans. Figure 5.5 shows the completed plan.

Now we will tackle the written part of the contract documents, which will completely define the conditions, components and installation methods that are required for our project.

Project specifications

The Construction Specifications Institute (CSI) has established a 17 division specifications standard that encompasses all of the work to be accomplished in a construction project. Within these divisions are separate sections that contain the specifications for work within the specific disciplines. Division 16000 is reserved for electrical work of all types. Section 16400 pertains to interior electrical systems, section 16500 covers lighting systems, and Section 16510 is for interior lighting systems, and is of particular interest to us. Each specification section is further broken down into three parts: Part 1 is the general conditions, which lists reference standards, special project requirements, project scope, and submittals required of the contractor; Part 2 is the products section, which lists the specific

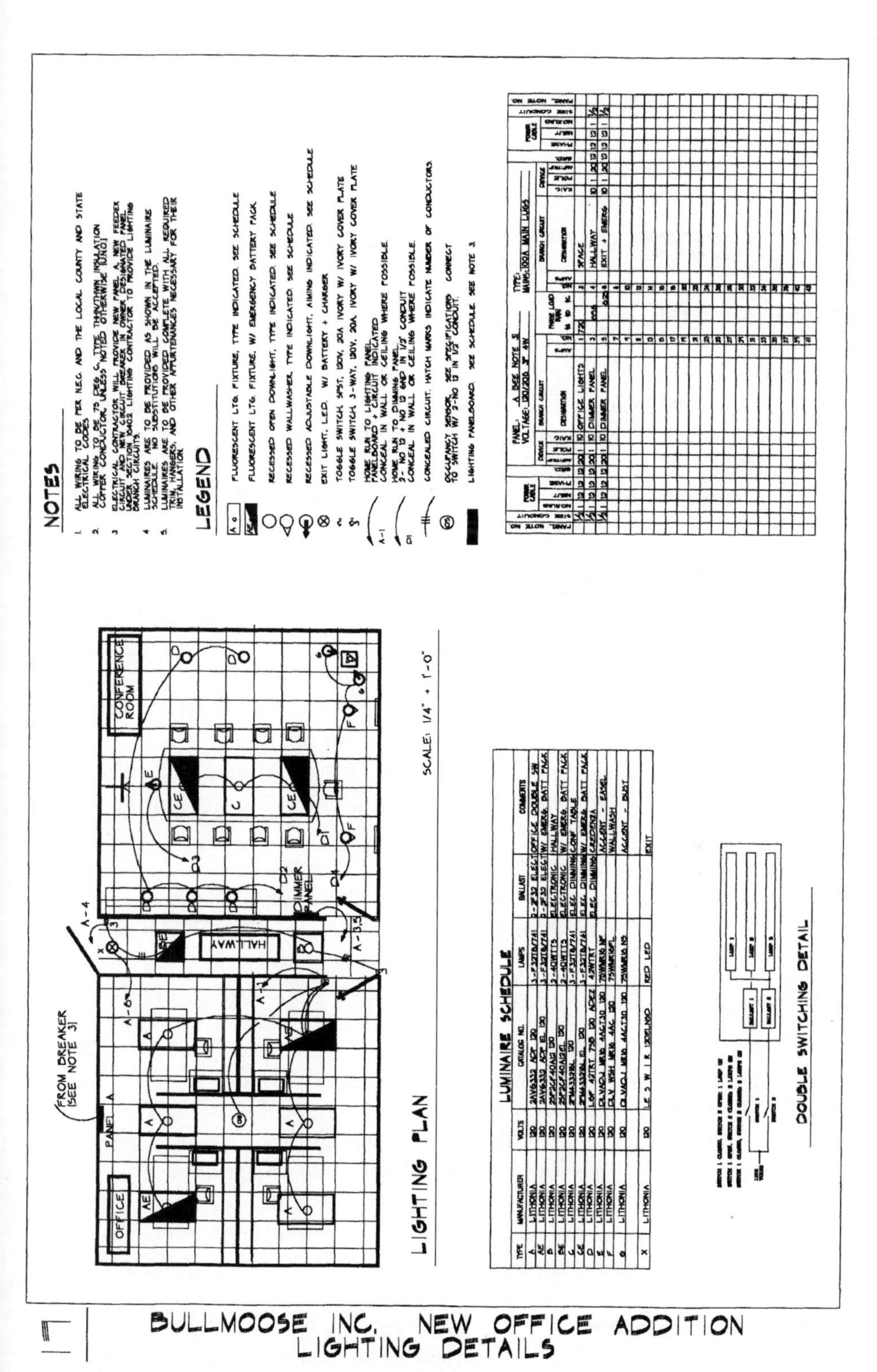

Figure 5.5. Bullmoose Office addition completed plan.

standards that the project components must meet; and Part 3 is the execution section, which spells out how the contractor is to install the components. What follows is a Section 16510 specification for our Bullmoose, Inc. project, with explanatory notes in italics.

SECTION 16510 – INTERIOR LIGHTING

Part 1. General

1.1. References

The publications listed below form a part of this specification to the extent referenced. The publications are referred to in the text by the basic designation only.

AMERICAN NATIONAL STANDARDS INSTITUTE (ANSI)

ANSI C82.1	(1992) Ballasts for Fluorescent Lamps
ANSI C82.2	(1989) Fluorescent Lamp Ballasts – Methods of Measurement
ANSI C82.11	(1993) High-frequency Fluorescent Lamp Ballasts

AMERICAN SOCIETY FOR TESTING AND MATERIALS (ASTM)

ASTM A 641	(1992) Zinc-coated (galvanized) Carbon Steel Wire

ILLUMINATING ENGINEERING SOCIETY OF NORTH AMERICA (IES)

LHBK	(2000) Lighting Handbook, Ninth Edition

NATIONAL FIRE PROTECTION ASSOCIATION (NFPA)

NFPA 70	(1999) National Electric Code
NFPA 101	(1994) Code for Safety to Life from Fire in Buildings and Structures

UNDERWRITERS LABORATORIES INC. (UL)

UL 924	(1995) Emergency Lighting and Power Equipment
UL 935	(1995) Fluorescent Lamp Ballasts
UL 1570	(1995) Fluorescent Lighting Fixtures
UL 1571	(1996) Incandescent Lighting Fixtures

All of the organizations listed above have put in a lot of effort and expense into developing exacting specifications for components that we have used in our design. By referencing these published standards, we incorporate the standards into our specifications.

1.2. Related requirements

Section 16050 'Basic Electrical Materials and Methods' applies to this section, with the modifications and additions specified herein.

Materials not considered to be lighting equipment are specified in Section 16402 'Interior Distribution System'.

The project electrical engineer normally prepares all division 16000 specifications, and references are made between sections, rather than rewriting the same clauses in every section. We will approach this section from the standpoint of the lighting designer who will integrate Section 16510 into the rest of the division 16000 sections prepared by others.

1.2.1. Work Included

The work included under these specifications consists of furnishing and installing all luminaires, controllers, wiring, conduit, and other necessary appurtenances to provide a fully functional lighting system as shown on the plans. The contractor shall provide all materials, labor, tools, and equipment necessary to complete the project.

1.3. Definitions

1.3.1. Average life (lamps)

The time after which 50% will have failed, and 50% will have survived under normal conditions.

1.3.2. Total Harmonic Distortion (THD)

The root mean square (RMS) of all the harmonic components divided by the total fundamental current.

The power supplies used in fluorescent electronic ballasts generate harmonics that can feed back into the electrical system and cause problems with sensitive electrical equipment. It is necessary for the lighting designer to specify a maximum allowable harmonic distortion for the ballasts.

1.4. Submittals

Other sections of the specification require the contractor to submit to the designer manufacturer's catalog sheets of the materials he proposes to use on the project. The designer then reviews the submittals, and either approves or rejects them. The contractor cannot install an item until he has an approved submittal for the item signed by the designer.

Submit the following in accordance with the section entitled 'Submittal Procedures'. Data, drawings, and reports shall employ the terminology, classifications, and/or methods prescribed by the IES LHBK, as applicable, for the lighting system specified.

1.4.1. Manufacturer's catalog data

a. Fluorescent luminaires;
b. fluorescent electronic ballasts;

c. fluorescent lamps;
d. incandescent luminaires;
e. low voltage incandescent lamps;
f. dimmer system;
g. exit signs;
h. emergency lighting equipment;
i. occupancy sensors.

This is the data that the contractor has to submit for approval. Switches, wire, panelboards, circuit breakers, and conduit are specified in Section 16402 by the electrical engineer. The lighting designer should coordinate closely with the electrical engineer to ensure that the components specified are compatible with the lighting system design

1.5. Electronic ballast warranty

Furnish the electronic ballast manufacturer's warranty. The warranty period shall not be less than 5 years from the date of manufacture of the ballast. Luminaires, including ballast assembly, shall be installed within 1 year after manufacture, leaving at least 4 years of the warranty in effect after installation.

Part 2. Products

2.1. Fluorescent luminaries

Luminaires shall comply with UL 1570. Fluorescent luminaires shall have electronic ballasts.

2.1.1. Fluorescent lamp electronic ballasts

The electronic ballast shall, as a minimum, meet the following criteria:

a. ballast shall comply with UL935, ANSI C82.11, and NFPA 70. Ballasts shall be designed for the number and wattage of the lamps shown on the plans. Ballasts are to operate on 120 V, AC;
b. power factor shall be 0.95 minimum;
c. ballast shall operate at a frequency of 20 000 Hz (minimum);
d. ballast shall have light regulation of plus or minus 10% lumen output with a plus or minus 10% input voltage regulation. Ballast shall have a maximum of 10% flicker;
e. ballast shall be UL listed Class P with a sound rating of 'A';
f. ballast shall have circuit diagrams and lamp connections displayed on the ballast enclosure. Ballasts shall operate lamps in a parallel configuration that permits the operation of the remaining lamps if one lamp fails;

g. ballasts shall operate in a rapid start mode;
h. ballasts shall have a full replacement warranty as specified in paragraph 1.5;
i. ballasts shall have a total harmonic distortion (THD) of 15%.

As a general rule of thumb, the lower the THD percentage, the larger the ballast power supply, and, as we remember from Chapter 2, the larger the current inrush when the lights are switched on. If you have a large number of luminaires with very low THD ballasts connected on a circuit, it is likely that the circuit breaker will trip when you turn on the switch; 15% THD is about as low as we need to go, except in special cases.

2.1.1.1. T-8 Electronic ballast

a. Ballasts shall have input wattage of 114 W maximum when operating four *F32T8* lamps.

2.1.1.2. T-8 Electronic dimming ballasts

a. Ballasts shall have the capability to continuously dim the output of its connected lamps from full output to 10% of full output. Dimming shall be accomplished by electronic circuitry responding to a low voltage (24 V) signal from a dimming system.

2.1.1.3. T5 Long twin tube ballasts

a. Ballasts shall have input wattage of 74 W maximum when operating two CF40T5 lamps.

2.1.1.4. Compact fluorescent electronic dimming ballast

a. Ballasts shall be solid state, dimming type, capable of continuously dimming the output of its connected lamps from full output to 10% of full output. Dimming shall be accomplished by electronic circuitry responding to a low voltage (24 V) signal from a dimming system.

2.1.2. Fluorescent Lamps

2.1.2.1. T-8 Lamps

T8 rapid start lamps shall be rated 32 W, 2800 initial lumens minimum, CRI of 72 minimum, color temperature of 4100 K, and an average rated life of 20 000 h.

2.1.2.2. T-5 Lamps

T5 long twin tube fluorescent lamps shall be rated 40 W, 3150 initial lumens minimum, CRI of 72 minimum, color temperature of 4100 K, and an average rated life of 20 000 h. Lamps shall be 22.6 in. maximum length, and shall have a 2G11 type base.

2.1.2.3. Compact fluorescent lamps

Compact fluorescent lamps shall be rated 26 W, 1800 initial lumens minimum, CRII of 72 minimum, color temperature of 4100 K, and an average rated life of 10 000 h. Lamps shall be double twin tube, with four pin base for dimming.

2.2. Incandescent luminaires

1. Incandescent luminaires are to be provided complete with transformers, and circuitry required for the operation of low voltage MR-16 lamps. The luminaires are to provide tool-less lamp adjustment in both the horizontal and vertical planes to allow aiming of the lamp. All required trim, lenses, hangers, etc., required for complete installation are to be provided with the luminaires.
2. MR-16 lamps are to be 75 W, with an average rated life of 4000 h.
   a. Narrow spot lamps are to have a beam spread of 10° maximum, with a center of beam candlepower rating of 10 000 candlepower. *This lamp will be used to highlight the bust of Commodore Bullmoose;*
   b. medium beam flood lamps are to have a beam spread of 24° maximum, with a center of beam candlepower rating of 3100 candlepower. *This lamp will be used to illuminate the presentation easel;*
   c. flood lamps are to have a beam spread of 36° maximum, with a center of beam candlepower rating of 1800 candlepower. *This lamp will be used to wall wash the portraits.*

2.3. Dimming system

2.3.1. Construction

The dimming system shall be housed in a steel frame with hinged faceplate. It shall be designed to fit in a standard five-gang backbox.

2.3.2. Functionality

The dimming system shall provide the following features:

a. single control to turn off all lighting;
b. master raise/lower control to raise or lower the lighting levels of all luminaires simultaneously;
c. channel raise/lower control to raise or lower the lighting level of luminaires connected to the individual channels;
d. preset controls for saving preset scenes;
e. adjustable fade time for preset scenes. Fade time will be adjustable from 0 to 45 seconds.

2.3.3. Capacities

The system shall be capable of dimming a minimum combined load of 2000 W. It shall be capable of dimming fluorescent electronic ballasts, compact fluorescent lamp ballasts, and low voltage incandescent loads. It shall provide a minimum of four separate dimming channels, and have a minimum of four preset scenes involving all channels. Each channel shall have capacity for:

a. 600 W of electronic ballast load;
b. 800 W of incandescent load.

2.4. Exit signs

Exit signs shall be self-powered LED type, conforming to UL924, NFPA 70, and NFPA 101.

2.4.1. Self-powered LED exit signs (battery backup)

Sign shall be provided with automatic transfer switch, battery, and automatic high/low trickle charger contained within the sign. Battery shall be sealed electrolyte type, and shall require no maintenance, including no additional water, for a period of not less than 5 years. The battery shall be capable of powering the sign for a minimum of 90 minutes.

2.5. Emergency lighting equipment

Emergency lighting equipment shall conform to UL934, NFPA 70, and NFPA 101.

2.5.1. Fluorescent emergency system

Shall be suitable for use with solid state ballasts. Each system shall contain an automatic transfer switch, battery, and automatic battery charger contained within an enclosure in a standard luminaire. The system shall include a test switch, operable from outside the fixture, and a pilot light, visible from outside the fixture. Battery is to be sealed electrolyte type, with capacity to provide power to one lamp at an output of 1100 lumens, for a period of 90 minutes. The battery shall require no maintenance, including no additional water, for a period of not less than 5 years.

2.6. Occupancy sensors

Occupancy sensors shall be UL listed. Sensor mounting type shall be ceiling surface. Detector shall be combination passive infrared/ultrasonic. Sensor shall have LED positive detection indicator, adjustable delayed off – time range between 30 seconds and 15 minutes, and sensitivity adjustment. Input shall be rated for 120 V, AC. Contacts shall be rated for 15 A minimum. Sensor shall have 'fail – on' function designed to keep the lights on if the sensor fails. Sensor shall have a manual override switch.

2.7. Support hangers for luminaires in suspended ceilings

2.7.1. Wires

Wires shall conform to ASTM A 641, class 3 soft temper, 12 gauge, zinc-coated finish.

Part 3. Execution

3.1. Installation

Set luminaires square, plumb, and level with ceiling and walls. Luminaires shall be in alignment with adjacent luminaires, and secure in accordance with manufacturer's directions and approved drawings. Installation shall meet the requirements of NFPA 70. Contractor shall obtain approval of the exact mounting for luminaires before commencing installation, and after coordinating with the other trades involved.

Lighting and HVAC (heating, ventilating, and air conditioning) diffusers compete for ceiling space in a grid ceiling. In this project, it is important that our luminaires be placed where we have shown them, and any conflicts should be worked out before the two contractors begin work.

Recessed luminaires may be supported from suspended ceiling system support tees when the ceiling system support wires are provided at a minimum of four wires per luminaire, and located not more than 6 in. from the corner of each luminaire. *In areas having seismic requirements, it will be necessary to provide an extra two wires per luminaire.* Provide support clips securely fastened to ceiling grid members, one at each corner of the luminaire. For luminaires smaller than the ceiling grid, provide a minimum of four wires per luminaire, located at the corners of the grid in which the luminaire is mounted. *In our case, this would be the recessed downlights in the conference room.* Do not support luminaires by ceiling acoustical panels. Where luminaires smaller than the ceiling grid are mounted in the center of the acoustical panel, support the luminaire with at least two 3/4 in. metal channels spanning, and secured to, the ceiling grid.

3.1.1. Exit signs and emergency lighting units

Wire exit signs and emergency lighting equipment on a separate circuit breaker in the lighting panelboard. Place a sticker beside this breaker which states 'EMERGENCY LIGHTING – DO NOT TURN OFF', and provide lockout.

3.1.2. Occupancy sensor placement

Locate the sensors in accordance with the manufacturer's recommendations to maximize energy savings and to avoid nuisance activation.

3.2. Field quality control

Upon completion of the installation, conduct an operating test to show that all components of the lighting system operate in accordance with this section.

And there we have it! We can send the contract documents to the architect *and* send him our invoice. Our work here is done, masked man. . . or is it?

When the project bids come in, the owner will probably suffer what is commonly known as 'price tag shock'. The cost for the project is higher than he had expected. He will huddle with the architect and the low bid contractor and try to come up with some ways to cut costs. The first item on the chopping block will be – you guessed it – the lighting system. The contractor will say that he can cut a lot of money out of the job if he is allowed to substitute generic luminaires for all that fancy stuff that you've specified. It's up to you to stand fast, and hold to your design. You have selected the luminaires through a careful process, and the system will function exactly as the architect has specified, at the lowest energy cost possible. You can enlist the support of the architect on this point. If your design is compromised, you can bet that, a year from now, nobody will remember the few dollars that were saved on the lighting, but they will curse the inferior system that they are having to endure. Your reputation as a lighting designer and the workers who occupy the space will both suffer if you allow the substitution of inappropriate fixtures into your design.

Enough of the soap box – now let's get on to those snappy exercises that we all know and love.

Exercises

1. What are the two uses that a contractor has for the contract documents?
2. What do the technical contract documents include?
3. Under what circumstances could the designer be required to pay for part of the project? What part would the designer be asked to pay for?
4. Why do we put a symbol legend on the drawings?

5. How do we identify wire and conduit sizes in our Bullmoose, Inc. project?
6. If you are designing a government project, funded by public money, how many manufacturers must you allow to bid on your project?
7. If you are designing a project that is funded by private money, how many manufacturers must you allow?
8. What is the purpose of the technical specifications?
9. How many divisions are there in a standard construction specification?
10. How many parts are there in each technical specification section?
11. Why do we list reference publications and materials standards in Part 1 of the technical specifications?
12. Why do we require the contractor to submit manufacturer's data on the products that he intends to use on our project?
13. What does the designer do with the submittals?
14. Why don't we specify 0% THD in our ballast specifications, since harmonics are undesirable in an electrical system?
15. Why do we require tool-less adjustment in our adjustable luminaires?
16. Why do we need a 'fail – on' function for an occupancy sensor?
17. Why does the lighting contractor need to coordinate with the HVAC contractor before starting work in the ceiling?
18. Why do we require an operational test after the installation is complete?

6 The second time around – retrofitting

The problem

Back in the 1970s, the 1980s, and even into the 1990s, open plan office spaces were all the rage, and the lighting design for these spaces was done largely by rule of thumb. The luminaire of choice was the 2 ft × 4 ft (0.6 m × 1.2 m) recessed fluorescent troffer with four F40T12/735 lamps, magnetic ballasts, and flat acrylic diffusers. These luminaires were laid out at 8 ft (2.4 m) on center throughout the space.

Given these parameters, the luminaire manufacturers could only compete on one basis – price. Manufacturers scrambled to cut manufacturing costs to the bone, and in doing so, the performance of the luminaires suffered. For the most part, this didn't matter much, because the spaces were so over-lit that even with a light loss factor (LLF) of 0.40, the design still produced plenty of light (and glare) for the workers in the spaces.

On the manufacturing side, the reflector panels in the luminaires were designed to minimize manufacturing costs, rather than to improve luminaire performance. Ballasts were built as cheaply as possible, at the expense of ballast factor. Even the acrylic diffusers suffered, by becoming thinner – some as thin as 0.10 in. (2.5 mm) – which reduced the prismatic diffusion pattern to randomly sized 'lumps' on the surface of the diffuser. All this served to give the luminaire a very poor efficiency.

As energy prices continued to rise, this luminaire inefficiency put an ever larger bite on the pocketbook of building owners, causing them, for the first time, to notice the lighting. Then, in 1992, along came EPACT 92, which outlawed the manufacture of cheap F40T12 lamps, and suddenly, building maintenance engineers couldn't find replacements for burned out lamps.

Lamp manufacturers came up with an alternative that met the requirements of EPACT 92, and which would operate on the installed 40 W magnetic ballasts. This was the F34T12 'Energy Saver' lamp, which was installed as the F40T12 lamps burned out. As we said earlier in this book, there is no free lunch. Lower wattage lamps in an already inefficient luminaire meant less light output, and in general, poor lighting. To make matters worse, the F34T12 lamps didn't provide the expected energy savings, because they operated inefficiently on the old 40 W ballasts. Many millions of square feet of existing office, hospital, and manufacturing space are presently operating under these conditions, using either aging F40T12 lamps, or the F34T12 lamps.

The solution

One solution to this problem would be to replace the old luminaires with new, efficient luminaires that use F32T8 lamps, and the much more efficient electronic ballasts. This is a very expensive option, and one that has a payback time (the time required for the energy savings to offset the installation costs) unacceptable to most building owners. A much more attractive option is to retrofit the existing luminaires to improve their efficiency. This means:

- improving the optics of the luminaires
- upgrading the ballasts of the luminaires
- installing F32T8 lamps and lampholders.

This will allow the existing luminaire housings and wiring to remain in place, while greatly improving luminaire performance and efficiency.

Just how do we go about retrofitting a luminaire? Let's take a look.

Improving luminaire optics

When the luminaires were built, the sheet metal that forms the housing of the luminaire was shaped to contain the ballast, and to mount the lamps, without much regard for optic performance. The resulting 'box' shape of the housing allowed light from the lamps to scatter within the housing, and most of the light from the upper half of the lamp was lost.

Reflectors are now being manufactured that redirect that light downward into the usable zone. These reflectors are available to accommodate the standard two, three, and four lamp configurations of the 2 ft × 4 ft (0.6 m × 1.2 m) troffer, and are designed to fit into existing housings with a minimum of extra hardware.

Typically, the reflectors are manufactured from material that has a 90–95% reflectivity, and that is formed to direct all available light out of the luminaire. When in place, reflectors alone can boost the efficiency of the older luminaires by 20–30%, with no increase in energy usage. It is common retrofit practice to replace four F40T12 lamps and two magnetic ballasts with two F32T8 lamps, one electronic ballast, and reflectors. This saves 100 W per luminaire in lamp energy, and about 30 W per luminaire in ballast loss.

The acrylic diffusers in the older troffers were often of poor quality, and as a result, scattered some of the usable light. Dirt accumulation and aging has reduced the transmissivity of the material, resulting in additional light loss. Some of the earlier diffusers were made of polycarbonate, which, as we saw in Chapter 2, yellows with age, and that further reduces the light output of the luminaire. Modern prismatic acrylic diffusers are manufactured with precisely formed prisms that serve to direct light downward as usable light. The diffuser is normally changed during the retrofit process. Now, what about the rest of the luminaire?

Upgrading the ballast

This is a simple one. We simply discard the old, inefficient magnetic ballasts and replace them with a new electronic ballast, don't we?

That's exactly what needs to be done, but it may not be quite that easy. The old ballast could contain PCBs (polychlorinated biphenyls), which we can't just throw in the local landfill, because PCBs have been determined to be hazardous. We first have to contact the ballast manufacturer and determine that the old ballasts do not contain PCBs. Then we discard them. If they do, they must be disposed of in a manner prescribed by the US Environmental Protection Agency (EPA). This costs a little more, but either way, we get rid of the inefficient magnetic ballasts and replace them with an electronic ballast. This gives us a big jump in efficiency since the magnetic ballasts can only operate two lamps – and therefore two ballasts are required for a four-lamp luminaire – which produces double the ballast loss. Now, what about upgrading the lamps?

Upgrading the lamps

Naturally, we want to go with the more efficient F32T8 lamps, but let's go one step further, and use 4100 K lamps, instead of the 3500 K 'cool white' F40T12 lamps that we're taking out. This will give us a little whiter light, and also the *perceived* impression of *more* light. This will help when we replace four F40T12 lamps in an inefficient luminaire with two F32T8 lamps in our improved luminaire. The actual footcandle level at the desktop may be the same, plus or minus a few footcandles, but the space will *seem* much better lighted. If you remember the opening paragraph of this book, psychology is an important part of lighting, and it will come into play here. As a bonus, we have reduced the amount of air conditioning required to remove the waste heat from the inefficient luminaires.

It's easy to see that we have reduced the energy consumption of the system. We have replaced four 40 W lamps with two 32 W lamps, and we have improved the ballast factor from around 0.6 to at least 0.8. Plus, we have reduced the air conditioning load. But that's not all. The maintenance costs for the retrofitted luminaires will be reduced by at least half, since we now have only two lamps and one ballast per luminaire to maintain. All of this contributes toward producing a low payback time, which makes retrofitting economically attractive to building owners. In addition, if you multiply all this by the millions of luminaires currently in service, you can see how retrofitting aging luminaires can greatly improve the national energy picture.

Let's take a concrete example, and see how it looks.

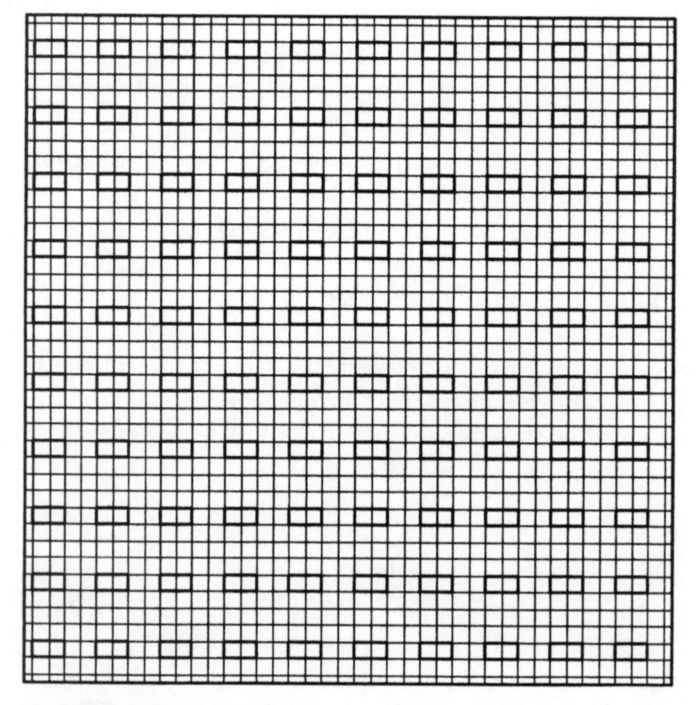

Figure 6.1. Office retrofit layout.

Let's say that we've been called in to survey a 80 ft × 80 ft open plan office space which is lighted by the typical four-lamp F40T12 luminaires, and we've been asked to show the payback for retrofitting the luminaires. The layout is shown in Figure 6.1.

We see that the luminaires are laid out on the typical 8 ft × 8 ft pattern, and that there are 100 luminaires. We can now compute the yearly energy savings for the system by calculating a 'before and

United Energy Associates, Inc.

PREPARED FOR: Example

Hilton Head SC.

FACILITY: RETROFIT

PREPARED BY: W.J. Fielder, P.E.

03/05/01

LIGHTING ENERGY AUDIT -- SAVINGS SUMMARY

SAVINGS SUMMARY

ANNUAL CASH SAVINGS	CAPITAL RECOVERY	RETURN ON INVESTMENT
$4,802	12.18 Months	98.48 %

CONVERSION SUMMARY

EFFECTIVE COST PER KWh: 0.0750
AVERAGE WEEKLY LIGHTING HOURS: 72
KWd RATE:

	BEFORE CONVERSION	AFTER CONVERSION
TOTAL FIXTURES:	100	100
TOTAL LAMPS:	400	200
TOTAL BALLASTS:	200	100
LIGHTING KWh PER MONTH:	5,990	1,841
LIGHTING KWd:	19	6
WATTAGE PER SQUARE FT:	3.000	0.922

ESTIMATED IMPROVEMENT COST AND SAVINGS

ESTIMATED COST:	$4,876	ANNUAL KWh SAVINGS:	$3,735
		ANNUAL KWd SAVINGS:	N/A, ,
NET IMPROVEMENT COST:	$4,876	ANNUAL HVAC SAVINGS:	$669
		ANNUAL MAINTENANCE SAVINGS:	$398
SAVINGS FOR FIVE YEARS:	$25,241	TOTAL ANNUAL SAVINGS:	$4,802
SAVINGS FOR TEN YEARS:	$53,799		

ANNUAL ENVIRONMENTAL SAVINGS	
POUNDS OF CARBON DIOXIDE:	39,836
POUNDS OF SULFUR DIOXIDE:	124,488
POUNDS OF NITRUS OXIDE:	59,754

Figure 6.2. ENLIGHTEN retrofit economic analysis (courtesy of United Energy Associates Inc.).

after' scenario, throw in the expected air conditioning savings, estimate the cost of retrofit, both labor and materials, estimate maintenance costs, run an economics analysis, and come up with an estimated payback time for the owner. If you do this by hand, it will probably take a day or two. A much easier approach is to use a packaged computer program to do the calculations for us. The program that we will use for this example is a proprietary program developed by United Energy Associates, Inc., of Winter Haven, Florida. The program is called ENLIGHTEN, and is available for purchase from United Energy Associates. Figure 6.2 shows the results of the ENLIGHTEN run for our example.

From Figure 6.2 it is seen that the capital recovery (or payback) time is 12.18 months, or a little over a year. After that, the savings to the building owner will be $4802.00 per year. This would represent a good value to the building owner, who would normally consider a payback time of under 3 years a sound investment.

For you environmentalists, you can also see that by reducing the energy consumption of the building, the amount of pollutants released into the air by generating plants has been reduced. In our case, the release of about 225 000 pounds per year of carbon dioxide, sulfur dioxide, and nitrous oxide has been avoided.

That's the retrofitting story. It's one of those rare cases where everybody wins. You as a designer should consider retrofitting as another valuable tool in your arsenal.

And that's about it for this book. Although I know you've grown to love them, you won't have any exercises to do for this chapter. Go forth and design!

Glossary

Accent Lighting: Directional lighting to emphasize a particular object or to draw attention to a part of the field of view.

Absorption: The dissipation of light within a surface or medium.

Accommodation: The process by which the eye changes focus from one distance to another.

Adaptation: The process by which the visual system becomes accustomed to more or less light than it was exposed to during an immediately preceding period. It results in a change in the sensitivity of the eye to light.

Air Fitting (air bonnet, air hood, air saddle, air box): A fitting that is mounted to an air handling luminaire and connects to the primary air duct by flexible ducting. It normally contains one or two volume controls.

Alternating Current (AC): Flow of electricity that cycles or alternates direction many times per second. The number of cycles per second is referred to as frequency. Most common frequency used in the US is 60 Hertz (Hz; cycles per second) in Europe, 50 Hz is the standard.

Ambient Lighting: General lighting, of the space (as opposed to task lighting or the lighting of the object one is looking at). It can be produced by direct lighting from recessed, surface or stem-mounted luminaires, or by indirect lighting that is wall or stem mounted, or by luminaires built into furniture or free standing.

Amperes (amps or A): The unit of measurement of electric current.

Baffle: An opaque or translucent element that serves to shield a light source from direct view at certain angles, or serves to absorb unwanted light.

Ballast: An auxiliary device used with fluorescent and HID lamps to provide the necessary starting voltage and to limit the current during operation.

'Batwing' Distribution: Candlepower distribution that serves to reduce glare and veil reflections by having its maximum output in the 30–60° zone from the vertical and with a candlepower at nadir (0°) being 65% or less than maximum candlepower. The shape is similar to a bat's wing. In fluorescent luminaires the batwing distribution is generally found only in the plane perpendicular to the lamps.

Beam Spread: The angle enclosed by candlepower distribution curve at the two lines that intersect the points where the candlepower is equal to 10% of its maximum.

Branch Circuit: An electrical circuit running from a service panel having its own overload protection device.

Brightness (Luminance): The degree of apparent lightness of a surface; its brilliancy; concentration of candlepower. Brightness is produced by either a self-luminous object, by light energy transmitted through objects or by reflection. Unit of measurement of brightness is the footlambert (fl).

Candela: The unit of measurement of luminous intensity of a light source in a given direction.

Candlepower: Luminous intensity expressed in candelas.

Candlepower Distribution Curve: A graphic presentation of the distribution of light intensity in a given plane of a lamp or luminaire. It is determined by photometric tests. The curve is generally polar, representing the variation of luminous intensity of a lamp or luminaire in a plane through the light center.

Capacitor: An electric energy storage device that when built into or wired to a ballast changes it from low to high power factor.

Cavity Ratio: A number indicating room cavity proportions calculated from length, width and height.

Ceiling Cavity Ratio: A numerical relationship of the vertical distance between luminaire mounting height and ceiling height to room width and length. It is used with the zonal cavity method of calculating average illumination levels.

Circuit Breaker: Resettable safety device to prevent excess current flow.

Class 'P' Ballast: Contains a thermal protective device that deactivates the ballast when the case reaches a certain critical temperature. The device resets automatically when the case temperature drops to a lower temperature.

Coefficient of Utilization (CU): A ratio representing the portion of light emitted by a luminaire in any particular installation that actually gets down to the work plane. The coefficient of utilization thus indicates the combined efficiency of the luminaire, room geometry and room finish reflectances.

Cold Cathode Lamp: An electric-discharge lamp whose mode of operation is that of a glow discharge.

Color Rendering Index (CRI): Measure of the degree of color shift objects undergo when illuminated by the light source as compared with the color of those same objects when illuminated by a reference source of comparable color temperature.

Color Temperature: The absolute temperature of a black body radiator having a chromaticity equal to that of the light source.

Cone Reflector: Parabolic reflector that directs light downward thereby eliminating brightness at high angles.

Contrast: The difference in brightness (luminance) of an object and its background.

Contrast Rendition Factor (CRF): The ratio of visual task contrast with a given lighting environment to the contrast with sphere illumination. Contrast measured under sphere illumination is defined as 1.00.

Cool Beam Lamps: Incandescent reflector lamps that use a special coating (dichronic interference filter) on the reflectorized portion of the bulb to allow infrared heat to pass out the back while reflecting only visible energy to the task, thereby providing a 'cool beam' of light.

Cut-off Luminaires: Outdoor luminaires that restrict all light output to below 85° from vertical.

Cut-off Angle (of a luminaire): The angle from the vertical at which a reflector, louver, or other shielding device cuts off direct visibility of a light source. It is the complementary angle of the shielding angle.

Dimming Ballast: Special fluorescent lamp ballast, which when used with a dimmer control, permits varying light output.

Direct Current (DC): Flow of electricity continuously in one direction from positive to negative.

Direct Glare: Glare resulting from high luminances or insufficiently shielded light sources in the field of view. It usually is associated with bright areas, such as luminaires, ceilings and windows that are outside the visual task or region being viewed.

Discharge Lamp: A lamp in which light (or radiant energy near the visible spectrum) is produced by the passage of an electric arc through a vapor or a gas.

Discomfort Glare: Glare producing discomfort. It does not necessarily interfere with visual performance or visibility.

Distribution Panel: Panelboard containing circuit breakers that distribute power to branch circuits.

Efficacy: See Lamp Efficacy.

Efficiency: See Luminaire Efficiency.

Equivalent Sphere Illumination (ESI): The level of sphere illumination that would produce task visibility equivalent to that produced by a specific lighting environment. Suppose a task at a given location and direction of view within a specific lighting system has 100 fc of illumination. Suppose this same task is now viewed under sphere lighting and the sphere lighting level is adjusted so that the task visibility is the same under the sphere lighting as it was under the lighting system. Suppose the lighting level at the task from the sphere lighting is 50 fc for equal visibility. Then the equivalent sphere illumination of the task under the lighting system would be 50 ESI fc.

'ER' (Elliptical Reflector): Lamp whose reflector focuses the light about 2 ft (61 cm) ahead of the bulb, reducing light loss when used in deep baffle downlights.

Extended Life Lamps: Incandescent lamps that have an average rated life of 2500 or more hours and reduced light output compared to standard general service lamps of the same wattage.

Floodlighting: A system designed for lighting a scene or object to a luminance greater than its surroundings. It may be for utility, advertising or decorative purposes.

Fluorescent Lamp: A low-pressure mercury electric-discharge lamp in which a fluorescing coating (phosphor) transforms some of the ultraviolet energy generated by the discharge into light.

Flux: Continuous flow of luminous energy.

Footcandle (fc): The unit of illuminance when the foot is taken as the unit of length. It is the illuminance on a surface one square foot in area on which there is a uniformly distributed flux of 1 lumen.

(Raw) Footcandles: Same as footcandles. This term is sometimes used in order to differentiate between ordinary footcandles and ESI footcandles. (Footcandles or Raw Footcandles refer only to the quantity of illumination. ESI footcandles refer to task visibility by considering both the quantity and quality of illumination.)

Foot Lambert: A unit of luminance of a perfectly diffusing surface emitting or reflecting light at the rate of 1 lumen per square foot.

Fuse: Replaceable safety device to prevent excess current flow.

General Lighting: See Ambient Lighting.

General Service Lamps: 'A' or 'PS' incandescent lamps.

Glare: The sensation produced by luminance within the visual field that is sufficiently greater than the luminance to which the eyes are adapted to cause annoyance, discomfort, or loss in visual performance and visibility.

Greenfield: Flexible metallic tubing for the protective enclosure of electric wires.

Grounding: Connection of electric components to earth for safety.

Group Relamping: Relamping of a group of luminaires at one time to reduce relamping labor costs.

Heat Extraction: The process of removing heat from a luminaire by passing return air through the lamp cavity.

High Intensity Discharge (HID) Lamp: A discharge lamp in which the light-producing arc is stabilized by wall temperature, and the arc tube has a bulb wall loading in excess of 3 W per square centimeter. HID lamps include groups of lamps known as mercury, metal halide, and high pressure sodium.

High Output Fluorescent Lamp: Operates at 800 or more milliamperes (mA) for higher light output than standard fluorescent lamp (430 mA).

High Pressure Sodium (HPS) Lamp: High intensity discharge (HID) lamp in which light is produced by radiation from sodium vapor. Includes clear and diffuse-coated lamps.

Incandescent Lamp: A lamp in which light is produced by a filament heated to incandescence by an electric current.

Instant Start Fluorescent Lamp: A fluorescent lamp designed for starting by a high voltage without preheating of the electrodes.

Inverse Square Law: The law stating that the illuminance at a point on a surface varies directly with the intensity of a point source, and inversely as the square of the distance between the source and the point.

Isolux Chart: A series of lines plotted on any appropriate set of coordinates, each line connecting all the points on a surface having the same illumination.

Joule Heating: Heating which occurs when an electric current is passed through a conductor having electrical resistance. The amount of heat generated is directly proportional to the resistance, and to the square of the current (Wjoule = I squared × R).

Junction Box: A metal box in which circuit wiring is spliced. It may also be used for mounting luminaires, switches or receptacles.

Kilowatt-hour (kWh): Unit of electrical energy, or power consumed over a period of time. kWh = watts/1000 × hours used.

Lamp: An artificial source of light (also a portable luminaire equipped with a cord and plug).

Lamp Efficacy: The ratio of lumens produced by a lamp to the watts consumed, expressed as lumens per watt (LPW).

Lamp Lumen Depreciation (LLD): Multiplier factor in illumination calculations for reduction in the light output of a lamp over a period of time.

Light: Radiant energy that is capable of exciting the retina and producing a visual sensation. The visible portion of the electromagnetic spectrum extends from about 380–770 nm.

Light Loss Factor (LLF): A factor used in calculating the level of illumination that takes into account such factors as dirt accumulation on luminaire and room surfaces, lamp depreciation, maintenance procedures and atmosphere conditions. See Maintenance Factor.

Light Output: Amount of light produced by a light source such as a lamp. The unit most commonly used to measure light output is the lumen.

Lens: Used in luminaires to redirect light into useful zones.

Local Lighting: Lighting designed to provide illuminance over a relatively small area or confined space without providing any significant general surrounding lighting.

Louver: A series of baffles used to shield a source from view at certain angles or to absorb unwanted light. The baffles are usually arranged in a geometric pattern.

Long Life Lamps: See Extended Life Lamps.

Low Pressure Sodium Lamp: A discharge lamp in which light is produced by radiation of sodium vapor at low pressure producing a single wavelength of visible energy, i.e. yellow.

Low Voltage Lamps: Incandescent lamps that operate at 6–12 V.

Lumen: The unit of luminous flux. It is the luminous flux emitted within a unit solid angle (1 steradian) by a point source having a uniform luminous intensity of 1 candela.

Luminaire: A complete lighting unit consisting of a lamp or lamps together with the parts designed to distribute the light, to position and protect the lamps and to connect the lamps to the power supply.

Luminaire Dirt Depreciation (LDD): The multiplier to be used in illuminance calculations to relate the initial illuminance provided by clean, new luminaires to the reduced illuminance that they will provide due to dirt collection on the luminaires at the time at which it is anticipated that cleaning procedures will be instituted.

Luminaire Efficiency: The ratio of luminous flux (lumens) emitted by a luminaire to that emitted by the lamp or lamps used therein.

Luminance: See *Brightness*. The amount of light reflected or transmitted by an object.

Lux: The metric unit of illuminance. One lux is one lumen per square meter (lm/m2).

Maintenance Factor (MF): A factor used in calculating illuminance after a given period of time and under given conditions. It takes into account temperature and voltage variations, dirt accumulation on luminaire and room surfaces, lamp depreciation, maintenance procedures and atmosphere conditions.

Matte Surface: A non-glossy dull surface, as opposed to a shiny (specular) surface. Light reflected from a matte surface is diffuse.

Mercury Lamp: A high intensity discharge (HID) lamp in which the major portion of the light is produced by radiation from mercury. Includes clear, phosphor-coated and self-ballasted lamps.

Metal Halide Lamp: A high intensity discharge (HID) lamp in which the major portion of the light is produced by radiation from mercury. Includes clear, phosphor-coated and self-ballasted lamps.

Nadir: Vertically downward directly below the luminaire or lamp; designated as 0°.

Outlet Box: See Junction Box.

'PAR' Lamps: Parabolic aluminized reflector lamps that offer excellent beam control, come in a variety of beam patterns from very narrow spot to wide flood, and can be used outdoors unprotected because they are made of 'hard' glass that can withstand adverse weather.

Parabolic Louvers: A grid of parabolic shaped baffles that redirects light downward and provides very low luminaire brightness.

Penumbra: The darkest part of a shadow.

Plug-in Wiring: Electrical distribution system which has quick-connect wiring connectors.

Point Method Lighting Calculation: A lighting design procedure for predetermining the illuminance at various points in a lighting installation by use of luminaire photometric data.

Polarization: The process by which the transverse vibrations of light waves are oriented in a specific plane. Polarization may be obtained by using either transmitting or reflecting media.

Power Factor: Ratio of: watts/volts × amperes. Power factor in lighting is primarily applicable to ballasts. Since volts and watts are usually fixed, amperes (or current) will go up as power factor goes down. This necessitates the use of larger wire sizes to carry the increased amount of current needed with low power factor (LPF) ballasts. The addition of a capacitor to a LPF ballast converts it to a HPF ballast.

Preheat Fluorescent Lamp: A fluorescent lamp designed for operation in a circuit requiring a manual or automatic starting switch to preheat the electrodes in order to start the arc.

'R' Lamps: Reflectorized lamps available in spot (clear face) and flood (frosted face).

Rapid Start Fluorescent Lamp: A fluorescent lamp designed for operation with a ballast that provides a low-voltage winding for preheating the electrodes and initiating the arc without a starting switch or the application of high voltage.

Raw Footcandles: See Footcandles.

Reflectance: Sometimes called reflectance factor. The ratio of reflected light to incident light (light falling on a surface). Reflectance is generally expressed in percent.

Reflected Glare: Glare resulting from specular reflections of high luminances in polished or glossy surfaces in the field of view. It usually is associated with reflections from within a visual task or areas in close proximity to the region being viewed.

Reflection: Light striking a surface is either absorbed, transmitted, or reflected. Reflected light is that which bounces off the surface, and it can be classified as specular or diffuse reflection. Specular reflection is characterized by light rays that strike and leave a surface at equal angles. Diffuse reflection leaves a surface in all directions.

Refraction: The process by which the direction of a ray of light changes as it passes obliquely from one medium to another in which its speed is different.

Romex: A 60°C cable comprised of an outer flexible plastic sheathing that contains two or more insulated wires, designated as type 'NM'. Romex is normally restricted to residential use.

Room Cavity Ratio (RCR): A numerical relationship of the vertical distance between work plane height and luminaire mounting height to room width and length. It is used with the zonal cavity method of calculating average illumination levels.

Rough Service Lamps: Incandescent lamps designed with extra filament supports to withstand bumps, shocks and vibrations with some loss in lumen output.

Service Entrance: Point at which power utility wires enter a building.

Shielding: An arrangement of light-controlling material to prevent direct view of the light source.

Shielding Angle (of a luminaire): The angle from the horizontal at which a light source *first* becomes visible. It is the complementary angle of the cut-off angle. In the case of a luminaire shielded by a reflector or parabolic cell louver, it is important to ascertain also the shielding angle to the reflected image of the light source, as this is often almost as bright as the source itself.

Silvered Bowl Lamps: Incandescent 'A' lamps with a silver finish inside the bowl portion of the bulb. Used for indirect lighting.

Spacing Ratio (SR): The ratio of the distance between luminaire centers to the height above the work plane. The maximum spacing ratio for a particular luminaire is determined from the candlepower distribution curve for that luminaire and, when multiplied by the mounting height above the work plane, gives the maximum spacing of luminaires at which even illumination will be provided.

Spectral Energy Distribution (SED) Curves: A plot of the level of energy at each wavelength of a light source.

Sphere Illumination: The illumination on a task from a source providing equal luminance in all directions about that task, such as an illuminated sphere with the task located at the center.

Task: That which is to be seen. The visual function to be performed.

Task Lighting: Lighting directed to a specific surface or area that provides illumination for visual tasks.

Three-way Lamps: Incandescent lamps that have two separately switched filaments permitting a choice of three levels of light.

Transformer: An AC device to raise or lower electric voltage.

Transmission: The passage of light through a material.

Tungsten Halogen Lamp: A gas-filled tungsten incandescent lamp containing a certain proportion of halogens.

Veiling Reflections: The reflections of light sources in the task that reduce the contrast between detail and background (e.g. between print and paper) thus imposing a 'veil' and decreasing task visibility.

Vibration Service Lamps: See Rough Service Lamps.

Visual Comfort Probability (VCP): A discomfort glare calculation that predicts the percent of observers positioned at a specific location, (usually 4 ft, or 1.2 m, in front of the center of the rear wall), who would be expected to judge a lighting condition to be comfortable. VCP rates the luminaire in its environment, taking into account such factors as illumination level, room dimensions and reflectances, luminaire type, size and light distribution, number and location of luminaires, and observer location and location and line of sight. The higher the VCP, the more comfortable the lighting environment. IES has established a value of 70 as the minimum acceptable VCP.

Visual Edge: The line on an isolux chart which has a value equal to 10% of the maximum illumination.

Visual Field: The field of view that can be perceived when the head and eyes are kept fixed.

Volt (V): The unit for measuring electric potential. It defines the force or pressure of electricity.

Wall Wash Lighting: A smooth even distribution of light over a wall.

Watt (W): The unit for measuring electric power. It defines the power or energy consumed by an electrical device. The cost of operating an electrical device is determined by the watts it consumes times the hours of use. It is related to volts and amps by the following formula: Watts = Volts × Amps.

Work Plane: The plane at which work is done, and at which illumination is specified and measured. Unless otherwise indicated, this is assumed to be a horizontal plane 30 in. (76 cm) above the floor.

Zonal Cavity Method Lighting Calculation: A lighting design procedure used for predetermining the relation between the number and types of lamps or luminaires, the room characteristics, and the average illuminance on the work plane. It takes into account both direct and reflected flux.

Index

(References to diagrams, tables and illustrations are in *italic*)